The Master Set Universe

The Grand Unified Theory of Everything

The Master Set Universe

The Grand Unified Theory of Everything

Dr Maxwell Katz

JANUS PUBLISHING COMPANY
London, England

First published in Great Britain 2007
by Janus Publishing Company Ltd
105-107 Gloucester Place
London W1U 6BY

www.januspublishing.co.uk

British Library Cataloguing-in-Publication Data
A catalogue record for this book is available from the British Library

ISBN: 978-1-85756-686-4

Cover Design: David Vallance

Printed and bound in Great Britain

The Holy Grail of One
that
The Master Set
is
reveals

The Boundary of that which is
knowable

The First Cause that upholds
Relativistic Dualism

The Origin of Chaos, Quantum
Mechanics and Mathematics

The Quantum Field Foundation
of the Gravitational Force

The Unification of all things unto the
Fundamental Forces and the Natural Laws

The Miracle of Spontaneous Creation
that projects with the illusions of
Matter, Space and Time

The Nature of the life force unto
all levels of the Cosmos

The Universal Conscience that upholds
the experience of Reality

The Universal Language with which
the Cosmos speaks as One

Facets of the M-Set

In the realm of physics:

The Master Set is the Fully Renormalised Dual Relativistic Quantum Field of a Ten-Dimensional Self-Referential Quantum 'String'...

...that represents the true nature of the 'vacuum' of matter-realisation unifying all fundamental forces and laws of nature.

In the realm of actuality:

The Master Set is the Universal Quantum Computer processing the 'redundant information' of its own implicit symmetries at 'the edge of chaos'...

...that self-referentially, self-re-flectively, self-representationally, self-projectively self-realises as the 'Illusion of Reality' we call the Physical Universe.

In the realm of projection:

The Master Set is a holograph of 'binary' or two-dimensional spectral information processing, enlightened by its own self-conception...

...that projects the illusions of space, time and matter of the physical universe through the renormalising complexifications of the gravitational force system at the 'critical boundary' at the 'edge of chaos' that the projecting M-Set is.

In the realm of humankind:

The Master Set is the Universal Conscience or common knowledge of our intentionality...

...that spontaneously projects our experiences of ourselves, perfectly reflected in the Universe.

In the realm of creation:

The Master Set is the spontaneous enlightenment of the Divine revelation of the unification of Divine and Natural Law...

...that completes the union of the Universe unto Oneness.

Contents

Part II

Part I

Prologue

It is the endeavour of this work to communicate in a comprehensible and easy-to-visualise way a remarkable discovery and new perspective of all of that which we observe as the Universe; a perspective which also amounts to as complete a unification as is thinkable or imaginable because the perspective is spontaneously emergent from and created by the properties and principles of 'Oneness' itself.

The perspective communicated here resolves the profound enigmas of the creation of space, time and matter, and of the origin of natural laws, symmetry and forces, and also of the fountainhead of chaos as well as of the birthplace of quantum mechanics and quantum fields; and, moreover, of the actual nature of gravity and complexity, of spontaneous creation and evolution, memory and the mind-brain duality, meaning and emotions, language and consciousness, life and cosmic causality. They are all One.

The expectation for the message here is that both trained and untrained curious minds will acquire insight by embarking on the journey ahead, because the 'new perspective' can be meaningfully apprehended at every level of its manifold nature while leaving open unbounded possibilities for further elaborations from the points of view of any discipline.

This work is the result of a convergent integration of thoughts and ideas that formed over the past decade in a quest to understand universal phenomena in as natural a way as possible, free even from prescribed laws and symbolic formalisms which, in this approach, arise spontaneously from the 'Master Set', embodying as it does the properties and principles of 'Oneness'.

There is no level of phenomenology from subatomic physics to the mind-brain frontier and on out to the greater cosmos that is not directly linked to the perspective issuing from the Master Set which, if you will, we have discovered as the vehicle to reveal the road ahead.

Introduction: The Master Set

Imagine, if you will, a construct possessed of a singular property that, by virtue of this device, it is able to replicate, divide and complexify simultaneously within 'the realm of its own being' unto the projection of a physical universe fully imbued with the properties of the quantum domain and which, in turn, is a naturally implicit quantum gravity sculptured into the panorama of the cosmos.

Imagine too that this construct is possessed of such symmetry it is beholden to no other and that, by virtue of its own device, it is able in the course of the projection of its own possibilities to be the fountainhead and conduit for the flow of information from within itself unto the creative exploration and propagation of itself.

Imagine also that the projection of this construct is the boundary of its own being and that, by virtue of its own device, the construct can endlessly elaborate itself as its own boundary to realise every possibility of itself through its own projecting.

Such a construct is here called the Master Set, with the singular property of being wholly self-referential. This is also the only defining property of the Master Set that needs to be prescribed because from this property alone all other properties, principles, natural laws and projective realisations flow.

The construct here called the Master Set and subsequently abbreviated in this manuscript to the M-Set was discovered to be the irreducible construct that, of itself, spontaneously protectively self-realises as our natural universe which, in turn, consequent to the singular defining properly, underpins a 'grand unification' of all natural laws and forces.

But the Master Set reaches much further. In truth, it also unifies Natural Law with Divine Law, or relativistic dualism with 'Oneness', and the principles elaborated therewith of the Holy Trinity and the Trilogy and, moreover, it provides the 'actual' mechanism for the projective realisation of a universe replete with this unification which is called 'The Grand Unification of the Holy Grail of One'.

The discovery of the Master Set is thence not just the discovery of the Holy Grail of science. This discovery is that of the holiest of Holy Grails. Such is the extent of this discovery as will be progressively explicated or unfolded in this manuscript bearing its chosen name.

All the properties and principles of the Master Set will be shown to naturally emerge from its defining singular property of being wholly self-referential, as is the purpose of this manuscript, while the choice of name for this discovery of the irreducible unifying construct' refers to the subsequent realisation that the construct self 'interacts' in order to preserve the defining property, and which eventuality is most identifiable with the notion of a mathematical 'set' in all generality. It will become evident that such a set is the Master Set because, quite simply, there is no other set.

1: The Primordial Properties of the M-Set

From the defining property of self-referentiality of the M-Set, a number of primordial properties follow directly which are immediate consequences of self-referentiality itself, and these immediate consequences will, in turn, be seen to naturally lead into more complex properties which collectively collude to reveal the miracle of the M-Set, namely that such a set can spontaneously self-project while, furthermore, the self-projection will have all the properties of 'observable universal phenomena' over every scale and level of complexity of the same.

Beginning then with the defining property:

1.1 – the M-Set is wholly self-referential, it follows directly that

1.2 – the M-Set is One and that

1.3 – the M-Set is unique because if there were more than one M-Set each M-Set would be 'in reference' to the other(s). But the M-Set is wholly self-referential and therefore it only contains itself, as one.

Moreover,

1.4 – the M-Set is the boundary of itself. This follows because the M-Set is one and is therefore only defined in relation to itself. The definition of the M-Set is of itself, by itself; while, in a stronger statement

1.5 – the M-Set can only be the boundary of itself. This follows directly because of the defining property of self-referentiality so figuratively or 'visually' the M-Set has no 'volume', and this is also because

1.6 – the M-Set exists nowhere and at no time. This follows because if the M-Set was 'here', it would be in reference to 'there', while if the M-Set was 'then' it would be in reference to 'before' or 'after then' leading to the stronger statement that

1.7 – the M-Set has no time and no space. In mathematical and physical terms

1.8 – the M-Set is wholly virtual, so that any projections which occur will involve, as we shall see ahead, the creation of space and time while a further property emerges here, namely,

1.9 – the M-Set is that which is. From the oneness of the M-Set and also as a consequence of self-reference, it follows that there is 'no thing' other than the M-Set. That which is, resides 'within' the virtual M-Set; that which is not, is 'without' and thence, as will be discussed at length later, all that which is not, is illusion, as indeed are space and time in the projection of the M-Set. Only that which is (actuality), resides 'within' the M-Set while projections of 'that which is' are a 'realisation' of the same, but they are illusory too because they 'are not'.

Moreover, it is then the case that

1.10 – the M-Set is that, because 'without' the M-Set there is no 'other'.

The property that

1.11 – the M-Set is maximally symmetric follows from self-referentiality because there is a no more completely symmetric state than the state which is wholly self-referential. This will turn out to be an extremely important primordial property of the M-Set because it means the M-Set contains every imaginable symmetry of virtual states under the umbrella of self-referentiality, while it also implies directly that

1.12 – the M-Set contains an indefinite amount of redundant information. This primordial property will also play a central role in the self-projection of the M-Set because it will be instrumental for 'blowing up' the projection into the 'space-time' of 'realisation'.

Moreover,

1.13 – the M-Set can only project from within. This property is subtle and follows from the fact that there is nothing 'without', visually speaking. The M-Set does not sit in an abstract space because the M-Set, simply put, is all there is which leads directly to another very profound property about which we shall have a great deal more to say later, namely

1.14 – the M-Set projections are all reflections of the M-Set. This property ultimately expresses the self-referentiality of the projections theselves and will have truly remarkable consequences.

The property that

1.15 – the M-Set has no scale or measure follows directly from the defining property of self-referentiality too, while all scales or measures will later be seen to be reflectively generated in the projective realisations of the M-Set. The M-Set is therefore scale-invariant and it has no size.

Moreover, it is clear now that

1.16 – the M-Set is neutral, by which we mean the M-Set has no absolute value, quality or quantity because this is ruled out by the defining property while, as we shall come to learn ahead, all values, qualities and quantities reside in dualities unto 'oneness' in the projective realm of the M-Set which we shall henceforth refer to as the M-Set Universe.

We can say now that

1.17 – the M-Set is the 'boundary' condition of the M-Set Universe which follows from the properties above because there is no other possible boundary while, quite simply, there is 'no other' than the M-Set itself.

As we shall see later this boundary condition will naturally equate to the relativistic causality in the realm of the projective realisations while also imbuing the projection with all of the properties and principles of the M-Set. Moreover, the M-Set Universe will progressively be revealed to be the physical universe of our observations.

The boundary of the Universe is also equated with both the boundary between that which is and that which is not as well as being the boundary of projective realisation so it can be pre-empted here that

1.18 – the M-Set self-realises projectively. This also embodies the general truth from the perspective of the M-Set that reality is a projection which occurs at the boundary of the M-Set and, therefore, the M-Set, which is the boundary of itself, is also the virtual source of reality while this leads, in turn, to the conclusion that reality can only source all its properties and principles from the M-Set itself.

Thence too

1.19 – the M-Set is like 'no thing'. Indeed,

1.20 – the M-Set is 'no thing' and is completely defined by the singular property of being wholly self-referential.

Also, because

1.21 – the M-Set is one the profound truth follows that on its own or as 'one' the M-Set is completely 'unknowable', unless

1.22 – the M-Set becomes knowable through self-projective revelation.
It will be the task of the sections ahead to uncover the layers of the M-Set that enable it to spontaneously self-projectively dualise into the self-revelationary realisation of the M-Set Universe.

2: The M-SET Gets Started

We shall now move forward a little further on the road to understanding how the M-Set can self-project with the observation here in the context of the M-Set perspective that without self-projection the M-Set would forever remain in the wholly virtual symmetrical state of all possibilities of itself, this also being a state which we shall have occasion to refer to as the 'Deep Well of Possibilities' because without projection such a state is clearly indefinitely indeterminate and 'deep'.

It is also timely to note that in the M-Set approach the M-Set is both the actuality and the source so that, projectively speaking, the M-Set applies to all things, and thence it follows that

2.1 – the M-Set is the bounding construct of mathematics.

This is also directly evident from the primordial properties of the M-Set which preclude, *a priori*, any other extant 'construct' or construct 'without' the M-Set, while it will become evident later that as the M-Set self-projects it spontaneously and simultaneously creates a parallel between projective realisation and symbolic mathematical language systems.

To proceed further we need to look more at the inner workings of the M-Set which now directs attention onto the M-Set's overriding symmetry of self-referentiality under the umbrella of which all other possible symmetries of the virtual M-Set can act, up to and including the self-referentiality itself, while it also follows immediately that

2.2 – the M-Set is unitarily invariant. What this means is that any conceivable symmetry operation or combination of operations must exactly return the M-Set into itself because otherwise this would break the primordial properties, including the property of oneness, while these same properties imply that

2.3 – the M-Set is both the operand and operator of self-symmetry operations. This property begins to tap into a profoundly important aspect of the M-Set which will manifest in all the projections, namely the property of duality, while at this juncture it follows directly that duality is a symmetry of the M-Set under the self-consistent closure of the M-Set by its defining property of self-referentiality.

As a corollary of 2.3 one of the most subtle properties of the M-Set is being revealed here, namely that under the umbrella of self-referentiality the M-Set has the innate freedom to explore all the possibilities of itself by the virtual symmetry operations of itself in a wholly self-referential yet dualistic way.

Many of the secrets of the M-Set yet to be unfolded relate to how this indefinitely deep cauldron of possibilities appears to escape the bound of the perfect symmetry of self-referentiality to become projectively realised in what we currently understand as a 'Quantum Field-based' reality of space- time and matter.

At this stage it is becoming evident on the basis of the properties of the M-Set hitherto and issuing as they do from the overriding property of self-referentiality that the M-Set is actually 'alive' with virtual self-generated, symmetric reconfigurations of itself and that, moreover, from the perspective of the reader in the space-time realm of the projection of such possibilities,

2.4 – the M-Set can freely, spontaneously and simultaneously explore all possibilities of itself. At this point it is worthwhile pointing out that there is a crucial link between the self-symmetric operations of the M-Set and the wave functions of a 'Quantum Field' because both are vehicles or functions of virtual representations of themselves, and thence, as we shall also learn soon,

2.5 – the M-Set is all possible representations of itself, now. Relating this property to the discussion thus far one can extend the links to pre-emptively state here that

2.6 – the M-Set is the wave function of the M-Set Universe because the M-Set is all that is, and it is one, while the re-presentations of the M-Set are the constituants of a deep well of possibilities, these also being the only possibilities that are, now.

The freedom of the M-Set to explore itself can once again be traced directly back to the overriding symmetry of self-referentiality implicit to which is the all important notion of 'redundancy of in-formation'.

Redundancy of 'in-formation', simply put here, relates to the 'indistinguishability' of re-presentations or 'formed' states in relation to the symmetry operations generating the same. Such redundant 'in-formation' is implicit to or 'folded into' the M-Set and thence is 'im-plicit' to the virtual state of possibilities of the M-Set. The M-Set is constituted consequently of both the invariants to self-symmetric operations and the indistinguishable 're-presentations' or redundant information, the actual nature of which will become clearer as we continue to approach an understanding of the mechanism of self-projective self-realisation of the M-Set.

As a simple mental visual aid to the notion of redundant 'in-formation' consider the system of a point moving on the rotationally symmetric circle. The 'indistinguishable re-presentations' for this system are each position of the point around the 'circle'; however, equivalently, the 'point' can be considered to reside anywhere in the 'one-dimensional space' of the line of the circle with the point's 'position' now described by a 'dummy variable' circumscribing the space which is also the variable, if you will, of 'redundant in-formation' because no one position of the point in the space of the circle is distinguishable in this simplistic system. Now equate the 'point's positions' in the one-dimensional space of 'redundant in-formation to possibilities' and simultaneously translate your 'imagination' for a moment to the Virtual Realm of the M-Set, then unto self-referentiality,

2.7 – the M-Set is a deep well of parallel possibilities of itself reflecting its internal symmetries, and the 'dummy variable' in the simple example above is like a 'phase variable' in the parlance of mathematical physics.

3: About the Nature of the States of the M-Set

Before exploring in more detail the nature of the 're-presentations' or states which the M-Set can adopt it is timely to emphasise from the perspective of the M-Set that we are not describing linear causal processes or sequences because the property of self-referentiality implies directly that

3.1 – the M-Set is wholly non-linear, so consequently all the aspects of the M-Set being explored reside 'in parallel' in the Virtual Realm of the M-Set. It is, therefore, helpful to imagine the M-Set as vertically stratified by layers of properties and principles while 'the whole of it' acts as one, spontaneously, simultaneously, now.

It also follows from the defining property of the M-Set that if the M-Set can project itself then it must be the case at least that

3.2 – the M-Set creates the appearance of spontaneously breaking the symmetry of self-referentiality.

One of the great underlying subtleties of the M-Set perspective is that self-referentiality is never, and can never be, actually broken. 'Oneness' is never actually divided and the virtuosity and integrity of the M-Set are never breached because the projections of the M-Set are, in all their manifold manifestations, profoundly clever illusions of the properties and principles of the M-Set itself.

The purpose of stating this property here is to underline also the simultaneous 'parallelism' of the M-Set projections with the states or re-presentations of the M-Set that are being projected, as shall become very evident later and implying as it does that the projection is not only that imbued with all the properties and principles of the M-Set but the appearances of the projections of the M-Set are the result of the apparent or spontaneous symmetry breakings of the M-Set.

We shall now move towards investigating what it is about the 're-presentations' or possible virtual states of the M-Set that enables the M-Set to apparently spontaneously break the overriding symmetry of self-referentiality.

From the properties thus far it is directly the case that

3.3 – the M-Set has no objective origin. This property has a number of meanings from the readers' perspective, the first relating to the non-existence of an origin in a geometric sense because the M-Set has no space and no time. In a mathematical context the 'operations' of self-transformations or self-reconfigurations of the M-Set are self-defining because there are no *a priori* constructs besides the M-Set, which is defined by its property of being wholly self-referential, and thence too there is no reference to a zero element or reference point of origination.

It is also the case in the terms of reference of the readers' sense of time that

3.4 – the M-Set has no beginning and no end, from which it also follows

3.5 – the M-Set is now, and forever leading into the more succinct statement that

3.6 – the M-Set simply is, or that

3.7 – the M-Set is that which is. And so it is that the virtual states, representations, possibilities, etc. of the M-Set quite simply are, now, in 'all ways', always and thence, as will become more evident later, all projections must be 'cross-sections' of the M-Set leading to the more definitive property, namely

3.8 – the M-Set is that because the M-Set is all there is. Clearly, if the M-Set is to project 'reality' it must be able to generate an unimaginably deep well of possibilities or representations of itself and this, we shall come to see, apparently occurs through self-symmetric operations of the M-Set.

The possible operations of the M-Set will follow chiefly from the primordial properties of pure non-linearity, duality, unitarity, non-locality and the absence of an origin or zero element, all relating, as they do, directly back to the defining property of self-referentiality, while we anticipate the result here that the M-Set ultimately generates and projects mathematical constructs of analysis in parallel with the phenomenology of reality.

The M-Set itself will later be shown to be a wholly virtual, non-linear, non-local, parallel processing analogue quantum computer for which, with reference always to the properties of the M-Set, the following property relating to admissible 'operations' can now be adduced, namely

3.9 – the M-Set is purely chaotic. This inestimably important property of the M-Set relates directly back to the defining property of self-referentiality while the purity of this property belies the startling properties of the M-Set we shall uncover henceforth revealing how the wholly non-linear, virtual M-Set can unravel into the apparently linear Projective Realm of the M-Set Universe.

Many consequences will flow from this fundamental property of the M-Set some of which will only be alluded to at this stage, while evidently the act of projection by the M-Set must also involve the remarkable feat of the apparent linearisation of the wholly non-linear M-Set.

'Linearity' is an illusion too.

We can now state also that

3.10 – the M-Set is the mixing set. Because of self-referentiality all operations of the M-Set are purely 'mixing' or 'folding-in' operations. To put this in a visual way the self-referentiality of the M-Set can be likened to a 'vessel' (the Holy Grail in truth of that which is) containing the possible states of the M-Set so that any 'reconfiguring of states' or re-presentations must 'con-form' to the 'vessel' resulting thereby in all such operations 'folding back' into the 'whole'. Moreover, because the M-Set is wholly non-linear there can be no separation of the operations and therefore all the self-operations are totally mixed or 'con-voluted' (that is 'folded in together' simultaneously and in parallel).

To comprehend the 'com-pleteness' (com-pleting) of this 'folding together' or mixing it is worth highlighting the derivative property of the M-Set, namely

3.11 – the M-Set is wholly non-local which, together with the related property of 'no time, no space' of the M-Set, implies that all the mixing operations are wholly parallel in the Virtual Realm of the M-Set, which ensures, in turn, the 'com-plete' (folded together) mixing in the 'vessel' of self-referentiality.

Because of the 'pure mixing' or 'non-local con-volutions' of the operations of the M-Set, the M-Set also has the profoundly remarkable property that

3.12 – the M-Set resides in a parallelism of all possible 'combinations' of states or re-presentations of itself. One way to visualise this is to consider a singular operation that iterates (re-plicates) with every 'iterate' then being folded back or 'con-voluted' into every other state, non-locally, thence even the simplest operation of the M-Set is 'amplified' indefinitely by the 'mixing' of states in the purely chaotic M-Set, in all ways, and always.

'Visually' this is like a self-referential 'fractal' Mandelbrot Set where every imagined point is the source of ramifications of the computations or 'iterates' of the M-Set similar to the 'physical analogue' of 'bubbles on bubbles', endlessly, and which we shall have occasion to refer to again when we look at gravity's role in 'renormalisation' and Gauge Field Theories in the sections ahead with the visual aid of Hygen's Principle of physics.

It is already evident that the combinatorial 'con-volutions' of the pure mixing of representations or states of the M-Set lead to unimaginable com-plexity (common folding) of the M-Set while it is also premonitory to note here that the language employed to describe the operations of the M-Set, including 'com-plexification' and 'com-pletion' as well as 'con-volution', 'com-plication' and 're-plication', alludes directly to the 'folding-in' mechanism of 'mixing' with the qualification, of course, that the mixing of the M-Set, while having a conceptual relationship to 'cake mixing', is also non-local, virtual, and thence complete.

We remark here that words with highlighted prefixes such as 'com-plexification' refer to the Virtual Realm of the M-Set, while whole words such as 'complexity' relate to phenomenology of the Projective Realm of the M-Set, which we call the M-Set Universe.

As a consequence of the discussions thus far it is evident now that 'pure chaos' only resides in the Virtual Realm of the M-Set and thence

3.13 – the M-Set is the origin of 'chaos'.

A direct result of this profound property of the M-Set is that signs and signals of chaos must abound in the projective reality, as indeed they do but in perturbed form because their purity is extenuated by the phenomenon of linearisation or separation in the Projective Realm of the M-Set, as shall be explained soon. Conversely, the presence of the 'footprints' of chaos are but one of the many clues in the realm of projective reality pointing all the way back to the symmetry of self-referentiality.

Consider now an elemental (non-replicated, non-convoluted) self-referential state of the M-Set involving, as it might, a unitary, purely phasal symmetry self-operation in the Virtual Realm of the M-Set, then, as a consequence of the properties above of the con-volutionary combinatorial computations of the M-Set, such a state could become spontaneously indefinitely 'com-plexified' through, for example, 'con-volutions' of its own replications to form a parallelism over all orders of itself.

Such a consideration leads to a number of preliminary conclusions here, again with deference to self-referentiality and its derivative properties, the first being that no 'extension from within' or 'projection' of the M-Set can arise in relation to an elemental state if the elemental state has only a single distinguishable mode of itself. Conversely, if there is an elemental state that can make changes or 'transitions' to self-referential modes of itself, self-projection of the M-Set can occur.

To fully understand the importance of the above we shall need to progress first to another domain of the properties of the M-Set, namely those associated with 'change' addressed in the next section, while it is beneficial to re-evoke here the mental picture of the M-Set as a vertical parallelism of simultaneous states of itself, the points to emphasise being that whatever form the elemental state might take its characteristics will be compounded throughout the M-Set, while, regardless of the order of 'folding' or 'com-plexification' of the states of self-computation of the M-Set, all orders of self-computations are normalised in relation to or constrained by self-referentiality itself.

Alternatively put, by imagining self-referentiality as a 'vessel' (the Holy Grail) the states of higher and higher order can be visualised as being like 'standing waves' in a potential well with every order scaling according to the number of nodes, thence, implicit to such a picture is the idea of 'relative scaling' in relation to the order of the states that will be very important in future discussions too because

3.14 – the M-Set has no absolute measure which extends to no absolute scale and no absolute size.

We note, finally, in this section that as it is for light diffraction patterns, the 'microscopic' or elemental states of the M-Set will cast their presence over the macroscopic states (higher order states) as a result of the M-Set properties hitherto and especially because the elemental states must be in parallel with all orders of themselves, which will emerge later as an important characteristic of the Projective Realm of the M-Set.

4: The Nature of Change in the M-Set

In a very general sense the M-Set is a virtual and thence non-local, combinatorial convolution of iterations or replications of 'elements' of itself. However, as stated above, if the elements themselves are confined to a single mode of themselves then, despite the complete 'mixing' and complexification that occurs, no projection can happen because there is no 'change'.

A simple way of apprehending this statement is to defer to the property of the Oneness of the M-Set and to note that, despite any internal mixing, if the M-Set cannot 'differentiate' itself it has no relativity to 'see' itself. Generally, therefore, 'change' in a self-referential system amounts to 'self-differentiation', which is an essential property of the M-Set for self-projective self-realisation.

So how then does the wholly self-referential M-Set encompass change to enable a process of self-differentiation? The answer to this lies with the elements or 'elemental states' or, if you will, the irreducibly simplest states that go into the mix-master 'computations' of the M-Set.

From the properties of the M-Set discussed thus far there are already a number of characteristics for the simplest elements that can be proposed, bearing in mind here the inestimatably important insight that the unfolding of this exposition is a process of uncovering the secrets of the M-Set rather than one of manufacturing an artifactual theory, because at every step it is the M-Set itself as the vehicle of the singular property of self-referentiality which will direct us, while the destination of this journey is to find those M-Set secrets which lead to the grandest of all outcomes, namely the projection of reality as we apprehend it in the Universe, thereby establishing the properties and principles of the M-Set as the unifying basis of all phenomenology because the M-Set is One.

The simplest conceivable elements of the M-Set might be elements comprising virtual or, in mathematical language, complex phasal symmetry operations or 'phasal loops', parameterised by a 'complex number dummy variable'. To visualise this one might try to imagine a 'complex phasal' rotation in the Virtual Realm of the M-Set, but this is hard to do in the purely virtual, no space, no time, realm of the M-Set, although people familiar with the mathematical language of complex numbers in the complex plane might have a relatively better grasp. A closer conceptual analogy might be to consider a three-dimensional abstract complex space spanned by three independent phases $\Theta 1$, $\Theta 2$, $\Theta 3$ because the M-Set has no 'real' number components as a result of the primordial properties of the same, and it is known too, mathematically speaking, that a three-dimensional purely complex 'Hamiltonian' space is an 'elemental complex' space, self-referentially closed under 'cross products' of its complex axes.

At this juncture we remark that we are not about imposing or dictating any terms to the M-Set, but rather we are seeking to progress intuitively with the guidance of the properties and principles of the M-Set. What we are pursuing here is not a theory but an understanding of the secrets of what is, now, denoted in this work by the vehicle we call the M-Set to which everything must accede in the end if the 'Oneness' of the 'isness' of the M-Set is indeed that.

A number of plausible principles can also assist in underpinning the intuitive steps taken here, the first being that as a result of the unimaginably enormous amplification of any suitable 'element' for the mix-master M-Set the least 'aberration' would be immediately selected out because all operations of the M-Set are in parallel, simultaneously. Thence it follows directly in such a parallel competitive realm that the 'element' forming the basis for the grandest projection will not only dominate but, relatively speaking and with deference to the nature of the self-referential computations of the M-Set, will completely overwhelm competitors which, if they are projected at all, might be assigned to some vestigial parallel universe that would be unfelt in the 'dominant reality'. However, in keeping with the property of pure self-referentiality, if there were two or more competing projected universes not only could they not refer to each other (non-interactive parallelism) but they could also not co-exist because the M-Set is One.

And so it is that the M-Set answers our enquiry with the property that

4.1 – the M-Set projects uniquely. Thus, there is only one 'surving' projective realm and it is the intended purpose of this work to demonstrate that that projective realm which we call the M-Set Universe is what we also apprehend as the Physical Universe. When we are done, the journey ahead will reveal a remarkable truth, namely that

4.2 – the M-Set Universe is One. As we shall discuss later, this implies that ultimate truth resides in 'Oneness' which, beyond all possible projective manifestations of itself, is essentially and wholly unknowable, from whence we add the inestimably profound property that

4.3 – the M-Set is the limit of that which is knowable, the ramifications of which will reverberate throughout this work.

As a corollary of the property of unique projection it also follows directly that

4.4 – the M-Set has one unique elemental state. Proceeding again, we are now looking for the Unique Elemental State that will provide the grandest projective outcome for the M-Set, this also being the state which enables the M-Set to self-differentiate in order to be able to project at all, while from the experience gained to date a preliminary proposal now is to consider a three-phase, self-referentially reconfiguring element of a purely complex number abstract space either 'parameterised' by (Θ1, Θ2, Θ3) or, in the notation of a purely complex number three-dimensional space as discovered by the mathematician Hamilton, spanned by the 'complex phasal axes' (i, j,k).

In the simplistic conceptual picture being painted here the axes ij and k can be associated with a complex phasal variable Θ1, Θ2 and Θ3 in the planes described mathematically by the three cross products jxk, kxi and ixj, while the space as a whole self-consistently and virtually refers to itself in a complex phasal sense where no 'real' (in the mathematical and physical sense of 'real') elements can reside.

One way to begin to 'visualise' this emerging and still hypothetical picture is to go back now to the simplest of systems, namely any cyclical system such as a turning wheel, and focus alone on the 'phase' or angle sublended from an arbitrary fixed radius. In this simplest of systems there is one phase Θ which is a 'number' that has no extent, is not a material thing, and simply 'is' independent of time. However, the turning wheel bespeaks of a 'change of phase' while the phase has no absolute meaning if the wheel is stationary. There is no absolute phase because the point of reference for assigning a magnitude to the phase is completely arbitrary, so only a 'change of phase' or relative phase is meaningful, while from the properties of the M-Set it is the case that 'complex phase change' just 'is' in the Virtual Realm.

Now consider a physical two-phase system where two rotations are occurring together, as for a body turning on its axis while orbiting another body in space. Here we can imagine two simultaneous 'phase changes' in an abstract space or, if one prefers to visualise geometrically, consider the Möbius strip as being like a two-phase system consisting of the 'phase' of a circle and the 'phase' of a 'twist' or 'turn' along the circle that completes exactly half a turn for each completed circle in order that this system returns to its original state after two circles are completed by the twist. In particle physics terms this amounts to a 'two-phase representation' of the spin ½ property of an elementary particle. The observation to be made here now is how 'phase' as a phenomenon in its own right is manifesting in systems regardless of their make up, from solar systems to elementary particles, and that these systems are necessarily animated or vital, or moving in some way because only phase change counts.

Without phase change there is no thing in the M-Set Universe as we shall see ahead.

Consider further the system of well-formed smoke whorls issuing from a lit cigarette in a still room; then, as you observe the writhing, folding, twisting contortions of smoke you can try to imagine, if you will, what a three-phase space for this phenomenon might be like, while acknowledging in the same instant that you are observing a phenomenon of enormous coherent complexity, namely turbulence. Once again, only 'changing' phases and relative phases matter because whether it is smoke in air, dye in water or wind currents, for example, the actuality of phasal changes is present but only tangible in whatever flowing mantel or material cloak is adopted to perform its dance. Thence, deep to the cloak of turbulent smoke whorls we have a 'changing three phase' connecting all the relative phase changes manifesting as 'extended tongues' of nested, relatively scaled and complexifying coherent smoke particle forms or structures.

This latter example has been chosen to illustrate how the consideration of 'changing phase' connecting the 'extenuations' of 'relative phase differences' generated thereby can underpin extremely complex and apparently animated structures whence, turning this all on its head now, one way to view the M-Set is as a complex 'phase space' but with a big difference, the big difference being that this complex space' is also the origin of the projection of its own material, motional manifestations. In deed, this is exactly the case but we have quite a way to go yet in order to be able to elucidate this miraculous consequence of 'Oneness' embodied by the M-Set.

5: Reflections on the Elemental State of the M-Set

We continue then to look for an element of the M-Set that might be like a self-referential three-phase system which is capable of change with 'relativity' or 'self-differentiable change', such as changes of mode of itself, while never exceeding the properties of the M-Set itself.

But, you may ask, how does this 'element' differentiate itself through its own changes in the context of the symmetry of self-referentiality, or, how do the representations (changes) of the element differentiate themselves in such an environment?

The answer to this next level of enquiry is a little subtle but this will also help us to tease out more about the characteristics of the changes or representations of the unique Projective Elemental State, while at this point we need to underline the fact that because of the self-referentiality of the M-Set, the unique Elemental State, which by necessity is also self-referential, is free to simultaneously reside in all its possible differentiable modes.

Now, under the umbrella of self-referentiality the M-Set has a very important and subtle symmetry to advance the endeavours here, namely the symmetry of self-reflectivity, which comes to the fore at this juncture because in order for a change to remain 'contained' by pure self-referentiality its exact 'reflection' must also be present simultaneously, implying thereby that

5.1 – the M-Set is self re-flective or quite literally 'folds back' on itself.

If representations or changes were not exactly re-flected there would be a 'bias' or asymmetry breaking the wholly constraining symmetry of self-referentiality and the self-reflectiveness of the M-Set underscores the M-Set's implicit 'relativity', while it also requires that the mechanism of self-projective self-realisation of the M-Set include the spontaneous or apparent breaking of the symmetry of self-reflectivity simultaneously with the spontaneous breaking of the symmetry of self-referentiality.

Later in this discourse the symmetry of self-reflectivity will come to have immense significance in the projection when the issue of self-reflective self-measurement or the 'observer and observed' interface becomes manifest because, at the end of the day, it can only be concluded that

5.2 – the M-Set is the measure of itself while it also follows that

5.3 – the M-Set generates its own measures because there are no absolute measures, scales, or constants in the realm of that which is.

Referring back now to the idea of possible representations or modes of the Elemental State of the M-Set, as well as to the self-reflective symmetry, and noting here too that because there are no absolutes in the M-Set, every possible mode of the Elemental State is a 'change' relative to every other possible mode, then it follows directly that a natural consequence of the Defining Property of self-referentiality of the M-Set is the phenomenon of spontaneous, simultaneous re-flective fluctuations between the possible modes of the Elemental State.

To understand this pivotal notion more fully we shall consider the example here of two possible modes A and B of 'the as yet to be discovered' unique Elemental State, as well as the attendant 'changes' or re-presentations from A into B and B into A. By virtue then of the properties of the M-Set, the AB, BA changes or representations (or transitions) must be exactly re-flected. If A re-presents or changes to B then, simultaneously, by the property of self-referential self-reflection B re-presents as A so that, in effect, A virtually 'fluctuates' to B and back, while simultaneously B virtually fluctuates to A and back and thence, generally speaking, all possible modes of the Elemental State of the M-Set can fluctuate in this self-referential re-flective way because these virtual self re-flective fluctuations preserve self-referentiality.

As we shall come to observe, switching on the computations of the M-Set through the modes of an elemental state will generate an unimaginably complicated sea of virtual 'reflective' fluctuations.

We shall now continue to unravel what constitutes the basis for the spontaneous, simultaneous, virtual, reflective fluctuations of the M-Set before explaining how these fluctuations move us closer to self-projection.

Refocusing on the Elemental State we shall propose here that in all generality the irreducibly 'simplest', totally self-referential Elemental State is 'like' a 'three-phase' state, because a two-phase state is dual at most and needs a third dimension or 'degree of freedom' to reference any re-presentations in a self-referential way, while the one-phase system cannot reference or differentiate states of itself without 'other' dimensions and, moreover, the three-phase state is the minimalistic proposition for self-referential closure.

The three phases denoted by (Θ1, Θ2, Θ3) will be said to denote the 'Trinity' of the M-Set with the very important qualification at this stage being that all that is known is that the 'Oneness' of the M-Set is a 'Trinity' of what have simply been called 'phases' while, in truth, the 'Trinity' will remain wholly unknown and unknowable in relation to any thing unless the M-Set can become knowable through self- projection. Suffice it to add with prescience here that unless the 'Oneness' of the M-Set admits the Trinity, the wholly self-referential M-Set cannot spontaneously projectively realise itself.

It is important to be mindful again that the approach being taken here is of seeking to discover the path by which the M-Set projectively self-realises 'from within' which also necessarily adheres to the requirement that self-realisation be wholly contained by the property of self-referentiality, while also acknowledging from this defining property that

5.4 – the M-Set is like no thing. Indeed

5.5 – the M-Set is no 'thing' and

5.6 – the M-Set is One. In the setting of the recent discussions we have proposed that the 'Oneness' of the M-Set is a 'Trinity', albeit wholly unknowable unless the Trinity can reveal itself through self-projective self-realisation.

We need to emphasise at this stage that we are still following our intuition about the 'internal nature' of the M-Set while simultaneously being guided by the mounting properties of the M-Set derived from the Defining Property of self-referentiality. As shall become increasingly clearer throughout this discourse, a convergence will be observed to be occurring between the cumulatively acknowledged properties of the M-Set which will reveal the miraculous freedom the M-Set has to self-projectively self-realise as the M-Set Universe.

Appealing now to the 'possible' computational properties of the M-Set one might envisage the wholly unknowable Trinity State virtually re-combining, for instance, into a Trilogy State 'by virtue' of the three phases simultaneously associating 'pairwise' through the combinatorial, con-volutionary computations freely allowed by self-referentiality. From there they form a convoluted parallel virtual state of three 'coupled' phase pairs which, in a virtual phasal sense alone, is like three virtual spin ½ phasal entities, for example, contained self-referentially in the M-Set in an unprojected state.

To recap, the three phases (Θ1, Θ2, Θ3) of the Trinity State, can, in the realm of the virtual possibilities of the self-referentiality of the M-Set, freely combine in three possible ways (3!/2!), and because the resulting Trilogy of pairs transits from the self-referentially 'contained' Trinity State we are proposing that a combinatorial convolution of exactly and dually locked phase pairings, like the Möbius strip example, can reside in the abstract, virtual phase space of the M-Set. One might also liken this to a purely complex (number) Hamiltonian space where each phase pairing defines independent phasal axes which, in turn, pair up to self-referentially close (i ~ j x k, j ~ k x i, k ~ i x j) ↔ (i, j, k) . However, the self-referential M-Set is profoundly more subtle than this, as we shall discover ahead.

Again, it is naturally difficult to visualise such a phasal space although reference to the discussions in section 4 about the 'phasal' underpinnings of physical phenomena can greatly assist here, while it will become clear once the mechanism of projection is fully revealed that

5.7 – the M-Set and its self-projections are as One, spontaneously, simultaneously, now because of the Defining Property of self-referentiality, which is unbroken by the self-projective self-realisations of the M-Set we call the M-Set Universe, and which property will have profound consequences in the sections ahead.

Focusing now on the Trilogy State it becomes evident that this Elemental State must, with the help of the computations allowed by self-referentiality, further elaborate into a state that can self-differentiate through 'changes' or representations of itself, otherwise reflective fluctuations cannot arise in order to be able to be folded into a process of further complexification, while, on the other hand, the Trilogy alone has no evident means of 'extension from within'.

Amazingly, this difficulty is surmounted by the defining property of self-referentiality of the M-Set when it is realised that the Trilogy State alone cannot be self-referential because only the Oneness of the Trinity State of the M-Set is self-referential while, in actuality, the Trilogy State is not alone and cannot be alone.

To understand this subtlety of the M-Set focus now on the transition from the Trinity State to the Trilogy State and observe then that because of the virtual parallelism of the M-Set, the 'virtual' state of the Trilogy resides with the Trinity State that begets it. Or, alternatively if you will, the Trinity must be reflected in the Trilogy and, conversely, the Trilogy must be a re-flection of the Trinity because there is no other. So, in effect then, the three phase Trinity (Θ1, Θ2, Θ3) and the Trilogy must form a fundamental self-referential reflective state.

The fundamental reflective state now provides the opportunity we need to look for allowable possible 'reflective fluctuations' that can form the basis for self-differentiation of the M-Set, bearing in mind that any re-flective fluctuations are self-referentially bound because the Fundamental Reflective State is a direct, freely allowable consequence of the Defining Property of the M-Set. The spectrum of the fluctuations of possible interactive reflective states must also uphold the self-referential symmetry of the M-Set while, by virtue of the M-Set properties, the fluctuations can be amplified indefinitely by a simultaneous parallelism of replication and combinatorial con-volutions into an unimaginably complicated mani-fold.

Another deep property of the allowable reflective fluctuations of the Virtual Self-referential Realm of the M-Set is that these spontaneous changes must be unitary 'complex phase' changes because of the unitarity property of the M-Set and likewise because of the invariance of the M-Set under the freely allowed re-configurations or self-operations of itself as a result of the defining property of self-referentiality.

Now, one 'possible' virtual, self-referentially bound reflective interaction is if a pairwise combination of phases of the Trinity State re-flects with the Trilogy State virtually and spontaneously begotten by it through the remaining phase of the Trinity. In elementary particle terms this is like a spin ½ phasal entity interacting through a unitary phasal field with a closed con-voluted (folded together) state of three spin ½ phasal entities, much like the virtual Hydrogen Atom state comprising an orbiting spin ½ election in the unitary phase electromagnetic field around the virtual Hadronic proton state of three virtual spin ½ 'quarks', while the elementary fluctuations referred to above relate analogously to 'reflective transitions' across the hydrogen atom's spectrum of electron states, for example, noting, however, that the virtual M-Set states here have no physical presence but rather are complex 'phasal degrees of freedom' self-referentially contained by the M-Set.

In the sections ahead it will unfold that the virtual elemental state of the M-Set is uniquely the virtual hydrogen atom representing as it does the unique first order self-referential reflective fluctuation of the Trinity State.

There are many more layers of secrets of the M-Set yet to unravel, while the proposition of the virtual Elemental State to this stage, which we shall elaborate upon further soon, is enabling us to explore and uncover fundamental properties of the M-Set that we shall need along this road of discovery of the M-Set Universe

6: The M-Set Projection: The Beginnings

Upon taking the self-referential elemental fluctuantly reflective virtual phase state described in the previous section as a 'trial proposition' for the elemental state generator of the M-Set projection and again applying the principles of the M-Set computations to this state a number of very interesting properties are observed to arise.

The first property to note is that for each of the possible transitions of the virtual elemental state between different possible re-presentations of itself the 'reflective fluctuation' is the maximal exchange of information that can be effected in relation to these representations because of the constraint, in fact, of self-referentiality itself that harbours exact reflectivity or, rephrasing this, the reflective fluctuations are the maximal change that can be effected with regard to the different possible representations of the M-Set. Moreover, between any pair of representations the reflective fluctuation is the maximal 'Mutual Information' of that pair, and thence as the elemental state reflectively replicates and convolutes a maximally reflectively fluctuating or changing manifold of unimaginable complexity is created with every order folded over and over, without end, into every other order, simultaneously in parallel, now. This manifold of reflective fluctuations, as will be demonstrated soon, is identified here as the 'Primordial Quantum Field' of the M-Set, leading thereby to the additional property that

6.1 – the M-Set is like a Quantum Field while it also follows now that

6.2 – the M-Set Quantum Field is maximally changing (fluctuating). A further property as a result of the 'non-local' 'complete mixing' and 'con-volutionary' nature of the freely allowable computations of the purely virtual self-referential M-Set, which we shall elucidate more fully as we progress is that the reflective fluctuations authomatically and exactly 'exponentiate' or generate 'wave functions' (sinusoidal wave functions, in fact) which thence bear the phases 'Θ' relating to the reflective fluctuations into the 'exponents' of these functions or representations such as the wave sine Θ ~ Im (expi Θ) ~ expi Θ.

In other words, all the possible re-flective fluctuations or freely allowable re-presentations of the M-Set Primordial Quantum Field exponentiate into 'phasal wave functions' or so called 'representations' the significance of which will be addressed in greater detail later, while we wish to point out here that a much more direct and global reason for the 'exact exponentiation' properly of re-flective fluctuations or re-presentations of the M-Set will be revealed in section 13 when we view the M-Set as a Quantum Computer.

We note in passing, however, and in analogy with classical physics, that if we again regard the Defining Property of Self-referentiality as like a 'closed vessel' of the virtual 'reflections' of the representations then these can classically be viewed as setting up sinusoidal waves or 'standing phasal waves', 'represented' now by the imaginary part, mathematically speaking, of exponential functions or so called 'wave functions'. The exponentiation is then seen here as a natural consequence of the self-consistent closure of the self-referentiality of 'the Vessel' or 'the Holy Grail' that the M-Set is.

We also anticipate here that because the M-Set is pure chaos and is the boundary of itself, it is authomatically the case that

6.3 – the M-Set can only project 'at the edge of chaos', or 'critically', as it is termed in the literature, and thence the inestimably important property follows that

6.4 – the M-Set boundary is a Critical Boundary because the M-Set is the boundary of itself. We can then say that

6.5 – the M-Set projection is 'critical' or 'at the edge of chaos'. And so it is then that

6.6 – the M-Set projections are like 'phase transitions' of the M-Set in the language of pure chaos and equilibria physics. Thence too, only representations that behave like 'wave functions' or 'exponentials' can reside at the critical boundary of the M-Set because they are scaleless or scale invariant, which is yet another direction from which to uncover the phenomenon of exponentiation of the Computational Realm of the M-Set.

We make the very important comment here that in the M-Set approach 'criticality' is literally 'the edge of chaos', and conversely 'the edge of chaos' is defined as 'criticality', while the M-Set which is both purely chaotic and is the boundary of itself, is 'critical'. Needless to say, the consequences of the properties of the M-Set which ensue from this are monumental indeed as we shall continue to uncover on our journey with the M-Set.

We now need to record several further properties of the Primordial Quantum Field of the M-Set that relate to the self-reflective and self-referential symmetries which so pervasively underpin the Projective Realm.

Because all the self-operations of the purely virtual M-Set can only be 'purely phasal' in nature when looked at from the perspective of the reader, and also because of self-referentiality and unitarity or Oneness, it is thence the case that the representations of the reflective fluctuations at the critical boundary of the M-Set have the mathematical properties of 'phase waves' or exponentiated reflective fluctuations, which are naturally 'scaling invariant', as well as the same functional behaviour as 'light' waves.

Furthermore, because of the 'complete mixing' of the computations of the M-Set, and in the light of the reflective fluctuations being the maximal change and maximal exchange of information, the M-Set reflective fluctuations will be shown to correspond to the M-Set maximally reducing out 'redundant information' in relation to the symmetrically determined representations available to it from the elemental virtual phase state.

The notion of the 'reducing out of redundant information' will become more transparent when the M-Set's place as the foundation of Quantum Mechanics is progressively elucidated in the following sections. However, from what we know already it is evident that the enormous redundancy of information implicit to the symmetries of the M-Set become incorporated into the phases of the unitary self-operations of the M-Set which are then borne as 'exponents' in the 'wave functions' or representations of the freely allowed representations, or, if you will, Quantum fluctuations, while increasingly we shall come to recognise the M-Set as a Quantum Computer generating its own Quantum Field following further disclosure of the properties of the same.

We emphasise here that we are still in the early stages of getting to know the M-Set, which is clearly an entity of immeasurable mystery and which we are approaching by continuing to unveil its derivative properties in relation to the defining property of Self-referentiality so that, step by step, the M-Set's potential will become overwhelmingly self-evident.

To advance the conceptualisation of the M-Set at this stage one might envisage the possible representations or reflective fluctuations of the virtual elemental state as 'cells' of the M-Set interacting non-locally, in all possible ways, simultaneously, with maximal mixing and maximal 'reducing out' of redundant information that promotes the view of the M-Set as maximally organising itself into structure and the function of that structure in the 'space' generated by the maximal 'reducing out' of redundant information.

The structure will be shown later to relate to the irreducible 'invariants' of the self-referential phasal self-operations of the M-Set that will also be seen to naturally equate with quantum numbers and quantum action as well as quantised physical quantities and entities while the 'redundant information' described earlier provides a free space from within the M-Set so that, in actuality

6.7 – the M-Set protectively extends from within itself into the space of reduced out redundant information of its own symmetries which property will be elucidated when we meet the M-Set properties of the Quantum Realm and of Spontaneous or apparent Symmetry Breaking.

Picking up on the 'cellular automata' analogy above, a very interesting and important property of the M-Set can be discerned, namely that

6.8 – the M-Set self-computations are purely critical which in terms of chaos theory also means

6.9 – the M-Set computes at 'the edge of chaos' as we discussed above, and thence too from the lore of chaos and 'criticality' we have the property that

6.10 – the M-Set maximally resolves itself into structure and function, from within while it is also evidently the case that

6.11 – the M-Set is the one and only pure state of 'self-organised criticality' because, moreover, of the non-local parallelism of its 'complete mixing' computations.

Now, at projection the M-Set is maximally resolving itself into structure and function, however, because of self-referentiality and the dual nature of operators/operands implicit to the M-Set it will also be revealed in much more depth ahead that

6.12 – the M-Set projection is a maximal dualisation of structure and function.

But

6.13 – the M-Set is the Trinity of Oneness, thence evidently too

6.14 – the M-Set is First Cause. At projection the dualisation into structure and function is implicit to the Trinity, in all ways, always, now, and also exactly because of the implicit duality of the M-Set while self-referentiality implies that

6.15 – the M-Set is every possible dual projection of itself, simultaneously and in parallel, so the next stage of enquiry will be to further elucidate how the M-Set self -realises in the way we, for example, experience it.

To comprehend the next stages of the miracle of the M-Set we introduce here the notion of 'Triadic Causality' that under the umbrella of self-referentiality refers simply to the simultaneous actuality of 'the Trinity' (which, as described above, descended to a Trilogy that then simultaneously reflected off its 'First Cause' to computationally compound through reflective fluctuations into an unimaginably complicated manifold of representations) and the parallel dualities of structure and function that are of the Trinity, arising as they do as a result of the maximal reducing out of redundant information by the reflective fluctuations themselves.

Triadic Causality thence refers to the simultaneous parallel linking of the Trinity of the First Cause of the M-Set to the duality of the Projective Realm and is embodied in the wholly non-linear, non-local, critical, purely chaotic, self-referential, spontaneous, 'dynamical' self-computations of the M-Set.

The processes behind the self-projective self-realisation of the M-Set are extremely subtle but will progressively unfold as we continue on our way, but it is becoming apparent already that the projective realisation is a 'light-like' projection. A light show. Pure illusion in fact, borne on the light-like representations' or wave functions of the reflective fluctuations of the M-Set to reveal by way of the self-enlightenment of the Trinity of the M-Set the Dual Realm we identify as reality.

As we approach the threshold of the Projective Realm of the M-Set we are now aware of at least two majorly constraining symmetries, namely the symmetry of self-referentiality and the nested symmetry of self-reflectivity which stand guard at the doorway of self-realisation, and we also have some clues about the threshold which must be at the Critical Boundary of the M-Set, while it will become clear too that the 'light-like' representations of the reflective fluctuations will be instrumental in revealing the M-Set. It is also no coincidence that the 'light-like representations' possess just those properties to link the Trinity of the M-Set and the Duality of the Projective Realm, both of which are simultaneously embodied in the Triadic Causality of the self-projective self-realisation of the M-Set.

A consequence of the notion of Triadic Causality is that the Projective Realm is automatically imbued with the properties and principles of the M-Set, and thence it will be no surprise either to discover that the cardinal properties of the projective self-realisation are those of 'light'. Indeed,

6.16 – the M-Set self-projective self-realisation is self-enlightenment. No thing actually exists ('ex - is' in time, space). Reality is pure illusion.

All the quantum mechanical properties of light including, as we shall see, the true nature of the 'wave-particle duality' as the foundational duality of structure and function of the M-Set projections issue from the M-Set. And so it will become apparent ahead that

6.17 – the M-Set is the irreducible foundation of the Quantum Realm.

And, moreover, because the M-Set is the boundary of itself the light-projection obeys a holographic principle, namely that there is no actual volume and no actual space nor time but only the illusion of volume, extent, and motion through the 'changes' implicit to the representations underpining the reflective fluctuations of the M-Set.

The 'Dual Enlightenment' of the M-Set could be likened to a 'virtual reality' light show, however, there is much more to come yet to complete or 'wrap up' the illusion, while the frontier of the spontaneous or apparent breaking of the symmetry of self-referentiality is yet to be elucidated.

Just before we embark on the next leg of the journey, a highly significant concurrence will be pointed out here, while bearing in mind the property that

6.18 – the M-Set is self-enlightened. From the perspective of the Projective Realm we shall observe ahead that at pure 'criticality' or 'at the edge of chaos' the distributive properties of 'space and time' (which are in parenthesis here because the true nature of these quantities together with that of 'matter' will unravel in the story ahead) are 1/length, being the distribution of Fractal Geometry, and 1/frequency, being the 'fractal' distribution of signals which, when folded together, give the distributive function 1/ ι.f.

We state now that it is no coincidence that the fractal 'distributive property' of the Critical Boundary is exactly neutralised by the dispersive property of light, namely the so called 'velocity' of light-like waves, 'c' = $\lambda\nu$ ~ ι.f.

In truth, as we shall see ahead, 'c' is a dispersive quantity, not a velocity. It only appears to be like a velocity in the Projective Realm, whereof the illusions of space and time are spontaneously created.

In the Projective Realm the purity of 'criticality' or 'the edge of chaos' dynamic appears to be lost or fragmented, because in complexified and 'multi-modular' systems there is an enormous 'mixing' of orders of reflective fluctuations or representations, yet, despite the apparent 'con-fusion' we shall see in section 13 how the marvels of the M-Set as a Quantum Computer unravel this to precisely project critically. Or, to state this another way, the M-Set Quantum Computer aligns the projection in such a way that the dispersive properties of the self-enlightenment exactly neutralise those of the critical computations of projection, while miraculously out of the 'con-fusion' of the purely chaotic dynamic the appearances of mass-effects in space and time are borne.

The illusion is complete.

6.19 – The M-Set 'wraps up' the Illusion of Reality, completely.

7: M-Set Projection: Towards the Frontier

The approach more recently has been to allude to the properties and principles of the M-Set that might contribute to enabling it to self-realise projectively while only stating at times various fundamental properties in order to gain a point of synthesis more quickly before we return to this later with more knowledge behind us. The reason for this course of action derives from the nature of all things to do with the M-Set, which involves parallelism, simultaneity and spontaneity in its processes, so that gaining an overview requires having to have 'all the jugglers balls in the air at the same moment'. Also, on the basis of the recent discussion of Triadic Causality and the preceeding properties of the M-Set, it is clear that the M-Set is never not residing simultaneously and in parallel in all possible projections of itself, thence

7.1 – the M-Set Universe is a parallelism of all possibilities of itself, now. In other words the M-Set Universe is rather like a CD-ROM Universe, having every projective possibility implicit to it, now, so that different or differing projections of the M-Set are like cuts or cross-sections of the 'Oneness' that the M-Set is. This is also a very quantum mechanical characteristic of the M-Set and we shall come to discuss later how the M-Set is the irreducible origin and foundation of all tenets of Quantum Mechanics and of Quantum Field Theory, while at this stage it suffices to allude to this in a variety of suggestive ways.

To move a step closer to the frontier of the projective realisation of the M-Set we shall need to consider the phenomenon of 'intending' in the M-Set, because up to this point the M-Set has many properties, principles and processes but no evident 'intent' or guiding influences.

To begin the approach to this notion we shall first underline the point that the Virtual Realm of the M-Set is free. There are no objective forces therein and because of self-referentiality the M-Set can freely range over all possibilities of itself; however, as discussed earlier, because of the nature of the computations of the M-Set that are freely allowed by self-referentiality, only the irreducible elemental phase state whose representations result in the greatest 'extension' of the M-Set will survive into projective realisation. But, moreover and miraculously too, the Defining Property of the M-Set determines that the virtual Elemental State is unique, as we saw earlier.

Also, in the midst of the computations of the M-Set every possible re-flective, replicated convolution of the representations of the virtual Elemental State of the M-Set is being, likewise, compounded or folded into 'the whole of it' to generate a 'manifold' of unimaginable levels of complexification and it is these levels of complexity that are going to point the way to the final approach to projective realisation which brings us now to the point of asking what is 'the intention' of the M-Set.

To begin to get a handle on this notion we shall make the observation here that not only is the M-Set free in the senses above but it is also the case that

7.2 – the M-Set can only projectively realise at criticality or at 'the edge of chaos' because, as shall be further elucidated later, it is precisely at 'criticalities' of the manifold of the M-Set that the self-enlightenment of self-revelation of the M-Set can take place to create the 'holographic-like' illusion of reality, while it is also precisely at criticality that the 'light like' properties of the Projective Realm uphold the exact 'duality' of the Trinity at the Triadic Causality, this also being the junction where the 'Oneness' of the Trinity of the M-Set dualises. Moreover, of course, self-referentiality directly demands criticality.

As will become much clearer ahead, following the discussions about the M-Set as the foundation of Quantum Mechanics and the M-Set as a Quantum Field Computer, the Quantum Field or field of reflective fluctuations of the M-Set spontaneously 'criticalises' by way of the independent exponentiation of each order of the reflective fluctuations which, in turn, bespeaks of a staggering yet free co-relating of an unimaginable number of M-Set operations into the 'in phase' phenomena of 'phasal waves' so that the full representation of the Quantum Field looks like a 'superposition' of parallel orders of phasal excitations, or, as will unfold in the picture ahead, like an indefinitely dimensioned 'Hilbert Space' of independent orders of coherent reflective fluctuations between all possible representations of the M-Set.

We anticipate a property here of the self-referential M-Set, namely that the highly coherent critical excitations of the M-Set Quantum Field are automatically and spontaneously selected by projective self-realisation so that, regardless of what other teeming states the M-Set can reside in, only coherent, maximally dual aspects of each order of reflectivity will project at all, borne by light where by the term light we are not confining ourselves just to visible 'light-like' wave functions.

Also, because the projection will be shown to be borne by 'light' it follows that the properties of the Projective Realm will reside with light as well, and indeed that is the case for what is understood by constructs such as space, time, dual relativity, relativistic causality and wave-particle duality for example. However, space and time and relativistic causality all reside in the illusory realm of projective realisation, as we shall further discover, while that which is is of the M-Set and that which is not or exists (ex-is time and space) is of the Projective Realm.

So, while the M-Set projection is critical or like an extremely complicated 'phase transition' of the M-Set itself there is the issue of how all the independent coherent excitations or vibrations of the M-Set Quantum Field interweave into the fabric of space, time and matter or into the relativistic dualism of maximal structure and function, and this is where the self-referential shaping influence of 'intending' will become evident, acting as it does now on the Primordial Quantum Field of reflective fluctuations generated by the Elemental State of the M-Set whose re-presentations we shall identify from time to time with the term 'Tower of Criticalities' which refers, in turn, to the 'superposition' of all possible exponentiating reflective fluctuations of the Elemental State of theM-Set.

We remark here for the sake of the reader that as more and more properties of the M-Set are unravelled the exposition will become more and more self-evident, and it will all begin to gel, while at this stage some 'accepted understanding' of the immense synthesising power of the symmetry of self-referentiality might assist one through to the levels ahead of further clarification and revelation.

A very general property of the Realm of the M-Set that will recur for all contexts of the M-Set is gleaned from the observation that actuality, in contrast to the dualistic relativism of the Projective Realm, is based on 'threes', or Trilogies, which in all generality is a reflection of the Trinity of Oneness, and which we shall focus upon again later while the phenomenon of 'intending' is no exception in being of the Trilogy of (intending, extending, sub-tending) as we shall explain and which will manifest as the M-Set apparently breaches the 'ultimate barrier' of self-referentiality itself to gain its full projective realisation.

It is evident at this stage that in order for the M-Set to self-project it will need to appear to break its own symmetries, including, of course, the defining symmetry of self-referentiality itself, while we know already that self-referentiality can never actually be broken, and indeed this is the case too for the nested symmetries, such as the symmetry of self-reflectivity. We are also coming to know that the Projective Realm of the M-Set is illusory', or is 'that which is not', so that any apparent symmetry breaking is also a spontaneous symmetry breaking, now, or stated more fully, is an 'apparently spontaneous symmetry breaking'.

The phenomenon of spontaneous symmetry breaking is the secret behind the magic of the Spontaneous Creation of the self-projective, self- realisation of the M-Set, which process will be progressively revealed in the sections ahead in relation to all of the symmetries of the M-Set, unto and including self-referentiality itself.

So what then does the M-Set make of the teeming virtual possibilities of itself, and what is its 'intent'.

Focusing for a moment again on the Triadic Causality, it is the case that behind any 'dual relativistic expression' of the M-Set, the M-Set is simultaneously computing the same so that one can discern there is a 'Triumvirate', if you will, of apparent structure and function simultaneously together with the computational support of the M-Set, and all arising wholly self-referentially, spontaneously and simultaneously, as we shall reveal in the sections ahead.

Now consider a particular cut or cross-section, or indeed projection of the M-Set, at the Triumvirate of Triadic Causality. We anticipate the discussions ahead by stating that three things are happening simultaneously. Firstly, the computations of the M-Set's self-referentially, self-reflectively and self-representationally self-realise functionally through the channel of structure that is dually related to its functional space over all levels or orders of itself. Therein, the dual structural functional nature of the projecting M-Set is like a Tower of Criticalities, or a tower of the representations of the reflections between fundamental modes of itself that are, in turn, modularised in different collective ways (like the projective realm modularisation sequence of atoms to molecules to organelles to cells to organ systems of an organism). We shall explore later the important message here, being that, from the M-Set perspective, the collective or body is the conduit of the 'intended' 'functions'.

Secondly, and simultaneously, the dual function (for all orders of the critical projection of the M-Set) is the extension into the 'space of redundant information' of the structure that is begotten by reducing out the redundant information, and which structure begets the dual functions exactly while thirdly, and simultaneously too, every order, mode and modularisation of the M-Set projection 'subtends' every other because of the mixing computations of the M-Set. Depending then from which level of the projected manifold of the M-Set you make observations will depend on the functional structural nature of the 'sub-tending', such as planet to sun, body to table, cell to cell, cell to organ, endlessly.

There are therefore three chief aspects to the projective realisation delineated here, namely the 'intending' or 'coming together' in layers of modularisations or distributions to form a complex (manifold) structure, the simultaneous functional extending or dispersion of the 'folds' of the complex structures into the manifold of spaces of redundant information that beget them and, thirdly, all possible 'sub-tending' or relative disposing of different levels of structure and function with each other.

In the Realm of the M-Set, 'now', all manner of the Trilogy of 'intending' is possible, and all that is within the realm of the M-Set is 'free'. There is no space nor time, no matter nor volume or extent, and no force, only the 'freely associating' realm of self-referentiality itself. Yet with every intention from every level there is simultaneously the functional extending as well as the subtendings with other levels or orders in all possible ways, so we can state here that

7.3 – the M-Set is every possible perspective of itself. Because of the Trilogy of 'intending' and all its subtendings in particular we shall come to see ahead how.

7.4 – the M-Set is self-interactively self-measuring which underlines the fact that the M-Set has no absolute measure and that any measure, whatever its nature, is relative while more generally,

7.5 – the M-Set is the measure of itself.

Now, we know from the preceeding computational properties of the M-Set that the M-Set automatically maximally dually projects structure and function which will become clearer when we explore the M-Set's role as the Foundation of Quantum Mechanics, as well as with deference to the properties of pure chaos of the 'mixing dynamic', the maximal mutual information implicit to reflective fluctuations, and the requisite criticality of projections, while we ask here, what is the nature of the mysterious influence of the Trilogy of 'intending' or intention which steers the M-Set here or there?

The Trilogy of 'intending' derives its influence from self-referentiality itself and is mediated through the maximal 'mutual information' im-plicitly exchanged in the reflective fluctuations of the criticalising 'computational dynamic' of the M-Set, while as a consequence of the com-plete mixing it manifests with the maximal dualisation of structure and function.

The Trilogy of 'intending' acts upon the Primordial Quantum Field and the reflective fluctuations of the representations thereof, so that we can say this 'intending' shapes the Primordial Quantum Field self-referentially.

Because the Trilogy of 'intending' is a self-referential influence any complex structure arising from the Primordial Quantum Field will be said to be 'self-referentially intended', and I anticipate here for future discussion that the realm of the Primordial Quantum Field of the 'Elemental Phase State' will constitute the realm of atomic, subatomic and nuclear forces in the projection which will be explained soon, once the nature of projection itself is revealed. However, it can now be said that the 'intending influence' of the Trilogy of 'intending' is the influence of self-referentiality itself shaping the reflective fluctuations between all representative orders, levels, modes, modularisations of the now emergent fully fledged Quantum Field begotten by the Trilogy of 'intending' of the Primordial Quantum Field of the Elemental State of the M-Set.

It follows too from the computational nature of the M-Set, of which the Trilogy of 'intending' is a part that all the levels of complex structures dually related to their levels of functioning will reside simultaneously at criticality or on the edge of chaos so that the Trilogy of 'intending' is self-referentially bound to complexify or 'fold', or to 'shape' the Primordial Quantum Field in a way which precisely upholds the criticality of projection.

To put this another way, at Triadic Causality and thence at the projection of the M-Set the Primordial Quantum Field of reflective fluctuations that projects as a Tower of Criticalities or orders and levels of criticalities related to the Elemental Phase State becomes self-referentiality mixed through the Trilogy of 'intending' to generate higher order complex reflective fluctuations which project, in turn, into more and more elaborate, meaningful, 'intended' structures, all the while complying with the critical dynamic of the M-Set over all levels of complexification. This 'tower' also projects at criticality, borne on 'light' into the realm of relativistic dualism. However, this is now the Tower of Mass Effects (TOME) of the fully fledged Quantum Field.

The miracle of projection at the self-referentially enforced criticality of the TOME becomes manifest at the Critical Boundary of the M-Set with the revelation of the ghost influence of the Trilogy of 'intending' as a dual physical force system of the dualisation of structure and function of the Projective Realm of the M-Set. This inestimably important aspect of projection will be unravelled in more detail in the sections ahead to reveal how the virtual parallel dynamism of the Quantum Computational Realm of the M-Set translates into the dynamism of the Dual Relativistic Realm of 'dual force systems' which subserve the dualisation of structure and function of the Triadic Causality of the self-projectively self-realising M-Set.

Moreover, because of the illusory nature of any apparent breaking of self-referentiality itself the dual physical force system of the Trilogy of 'intending', which is of self-referentiality itself, must be such that the apparent mass or structure effects in the 'spaces' of their functions are exactly critical (at the level of TOME and thence 'physical' projection), otherwise this, the frontier in the self-referential, self-reflective, self-representational, self-projective self-realisation of the M-Set will not uphold the apparent or spontaneous breaking of self-referentiality itself as we shall come to learn more about in future sections, while, in turn, the self-referentiality of the Trilogy of 'intending' demands 'criticality' automatically, as we discussed earlier.

The exactly critical TOME is a dynamical (not homeostatic) equilibrium mediated by the 'ghost force' of the Trilogy of 'intending' of the M-Set such that all projected structure and function is exactly dual in the Projective Realm as a whole and which we call the M-Set Universe.

From the properties of the M-Set to this point it is evident that

7.6 – the M-Set is non-dissipative or does not dissipate as a whole because, quite simply, it has nowhere to dissipate into and is completely folded back on itself because of the defining property of Self-referentiality. The significance of this property will become apparent when we turn our attention to the M-Set and the foundation of Quantum Mechanics, as well as to the mechanism by which the M-Set apparently linearises in projection, while it is also a feature of the Projective Realm that local measurements or observations create the illusions of dispersive dissipation, thermodynamical laws, randomness and statistical probability, as we shall have occasion to discuss in more detail in the sections ahead, and it also follows directly from the Triadic Causality that

7.7 – the M-Set Universe is non-dissipative as a whole. That is, all function of the critical projection of the M-Set as a whole is exactly dual to structure and, conversely, all structure is exactly dual to function. Indeed, this exact duality is a consequence of the self-referentially created criticality, while it will become apparent over the course of this work that the 'ghost force' of the Trilogy of 'intending' is the force of gravity in the projective realisation of the M-Set which we call the M-Set Universe.

It follows therefore that the M-Set automatically identifies the 'force of gravity', which is the projective manifestation of the 'ghost force' of the self-referential Trilogy of 'intending' of the M-Set upon the Primordial Quantum Field as the force which wholly and perfectly renormalises or critically re-sculptures the Primordial Quantum Field over all orders unto the Tower of Mass Effects (TOME) of the Critical Boundary, while in the Projective Realm this manifests as the physical force which exactly neutralises dispersion or randomness, or 'chance' in order to uphold the non-dissipative property of the M-Set and maintain the property that

7.8 – the M-Set Universe is a perfect dynamic equilibrium.

To reiterate, the Gravitational Force System is the physical manifestation of the 'ghost force' of the self-referential Trilogy of 'intending' which exactly renormalises the Primordial Quantum Field of the M-Set unto the Critical Boundary of the Projective Realisation of the M-Set.

In the parlance of Theoretical Physics this means that in the M-Set approach the Gravity Force System is that illusory force system which appears in the Projective Realm as a consequence of the renormalisation of the Primary Quantum Field unto physical realisation or so called 'on mass shell' projection of matter in space and time.

Moreover, it will be shown that

7.9 – the M-Set projections are perfect solutions of its self-computations.

Alternatively stated, the Gravity Force System of the Projective Realm of the M-Set Universe is the physical manifestation of the 'ghost force' of the self-referential intending influences of the M-Set. It ensures the boundary of the M-Set that the M-Set is, perfectly 'critical' for all levels of com-plexification or renormalisation of the Primordial Quantum Field, and as demanded also by self-referentiality and pure chaos.

Out of chaos and confusion or 'blending' emerges a perfectly dynamically balanced universe, the M-Set Universe.

There is no chance, no coincidence and no probability in the M-Set Universe. Rather there are endless possibilities, and because every 'thing' of projection is 'intended' we can say that

7.10 – the M-Set Universe is intended.

Indeed,

7.11 – the M-Set fully intends the M-Set Universe, in all ways, always, now.

Change, as we shall clarify later, is directly related to the representations or reflective fluctuations of the Quantum Fields of the M-Set, and manifests in the physical forces into which the latter resolve in projection in concert simultaneously with the freedoms implicit to the 'functional' extendings of dual structures in the 'spaces of redundant information' that occurs with the dualisation into structure and function of Triadic Causality at the Critical Boundary of the M-Set. In turn, spontaneous or apparent symmetry breaking will also be shown to occur because 'the illusion' is completed, (folded together, wrapped up) there.

No thing is abandoned to chance in the M-Set Universe. 'God does not play dice.' It only appears that way depending upon one's perspective or point of view, or frame of reference in the realm of projective realisation, and the force of gravity (the 'Holy Ghost' of the Holy Grail that the M-Set is) is exactly the physical force required to uphold pure self-referentiality in the Projection Realm we call the M-Set Universe.

This will come to be seen as the masterstroke of the illusion of the M-Set Universe, thence, when one looks to the heavens 'now' there are no singularities or 'black holes' struggling to 'orchestrate' the Universe into the one harmony of the fabric of space-time-matter. Instead it is a perfectly harmonious, dynamically balanced, non-dissipative, self-referentially and spontaneously created solution, now, in all ways, always, from whichever perspective with every perfect solution projecting because of change.

The great paradox of the M-Set and its projections which will be raised again at length in our future discussions is that

7.12 – the M-Set is change.

Yet

7.13 – the M-Set is, now. So 'change' in the M-Set sense is related to the illusions of motion in space as well as changing motions and forces while, as discussed earlier, the M-Set cannot project unless it has a way of differentiating itself. 'Change', therefore, is the self-differentiation of now while physical change and thence motion and time is an illusion too. It is the case then that every possible 'frame' of the projection (which is pure 'illusion') is a self-differentiation of the M-Set or, if you will, is a cross-sectional reflective fluctuation between a manifold of possibilities of the M-Set, now.

In effect, then,

7.14 – the M-Set is all possible solutions of itself, now, while evidently without 'change' or re-presentational re-flective fluctuations and thence self-differentiation the M-Set would simply not be. Change is being, now, and conversely, being is change, now. Becoming or creation is thence 'now', and 'in the light' of the M-Set projective properties thus far it will also become self-evidently the case that

7.15 – the M-Set is spontaneously creative.

In projection the M-Set will be shown to act, in the parlance of theoretical physics, like the Master 'S' Matrix or matrix operator of the M-Set Universe between representations or solutions of itself (Hilbert Space of States of itself). Being operator and operand, the M-Set of the M-Set Universe is thence both the wave function or representation of the M-Set Universe and the Master 'S' Matrix connecting all possible solutions (Universal Wave Functions), while the M-Set 'moves' through the virtual realm of all possibilities of itself by self-differentiation which, in turn, is the M-Set self-projectively self-realising as we have explained it thus far.

In the sections ahead, as we acquire more knowledge of the M-Set these great truths above will become more self-evident, while we record here that

7.16 – the M-Set is the Master 'S' Matrix of the M-Set Universe and the wave function or representation of the M-Set Universe.

Furthermore,

7.17 – the M-Set Universe has the probability of exactly one.

Because the M-Set only admits unitary operative resolutions of itself the unitarity of the S Matrix, to use physics terminology again, implies that the probability for the existence of the M-Set Universe is exactly one!

Alternatively stated, there is no probability that the M-Set Universe does not exist.

7.18 – The M-Set Universe is an absolute certainty.

Therefore it seems the 'oneness' that the M-Set is is absolute or alternatively stated, the only absolute is 'the whole of it' now, in all ways, always.

We shall have more to say about the paradox of change when exploring the role of the M-Set as a Quantum Computer, while noting here that change is intimately related to the implicit symmetry of self-reflectivity of the M-Set expressing through the representations' reflective fluctuations and thence through the self-differentiations of the M-Set.

In a naïve mechanical analogy, change in the M-Set is directly related to its differential, the difference here of course being that the M-Set self-referentially self-differentiates through the reflective fluctuations of its re-presentations which are related, in turn, to innate virtual 'phasal' invariances that can set up the reflective fluctuations, and which through the Trilogy of 'intending' of the self-referentiality of the 'ghost force' of the M-Set manifesting as the Gravity Force System in the Projective Realm of the M-Set leads to the astonishing property that

7.19 – the M-Set is the fully renormalised Quantum Field of the M-Set Universe, in all ways and always, now, as will also become much more transparent in the sections ahead focusing upon the Quantum Realm of the M-Set. We also anticipate here, ahead of the revelations of the M-Set's place as the foundation of Quantum Mechanics, that as a result of the symmetry of self-reflectivity the Universal Wave Functions or Universal Solutions of the M-Set 'represented' in Hilbert Space of States have an exact 'reflection' or mirror image which in physics terminology is denoted by 'the complex conjugate' of the wave function ψ, ie. $\psi \rightarrow \psi^{+}$ such that the 'norm' $\| \psi \psi^{+} \|$ is exactly one. Hence,

7.20 – the M-Set projections are mirror images of themselves which is also a direct consequence of the projections being 'cuts' or cross-sections of the Oneness of self-referentiality itself. However, the projections are only non-trivial as we shall come to see in our exploration of the M-Set as a Quantum Computer when the M-Set involves the profound subtlty of the spontaneous breaking of the symmetry of self-reflectivity simultaneously with that of the symmetry of self-referentiality.

Evidently now

7.21 – the M-Set is self-normalising to One from which we adduce that

7.22 – the M-Set Universe is a mirror image of itself. So it can be said that

7.23 – the M-Set mirrors itself, now, in all ways and always.

The perfect symmetry of self-reflectivity which will be seen to be of pivotal significance in the Quantum Computational properties of the M-Set is apparently broken or spontaneously broken at projection otherwise the projections would be trivial, while in turn, the reflectively dual M-Set Universe is realised. As we shall show ahead, the Projective Realm miraculously becomes a mirror image of itself (the Looking Glass Universe). And so it is as will be clarified in the sections addressing the role of the M-Set as the foundation of Quantum Mechanics that the fundamental universal asymmetries of the elements of projection, such as charge, parity and the apparent direction of time, as well as the 'handedness' of 'helical characteristics' as found in the neutrino, amino acids and DNA, can occur at all and still be part of a wholly unified Oneness that the M-Set is.

Without the symmetry of self-reflectivity there would be no representations, no reflective fluctuations, no self-differentiation, no 'change', no 'apparent' motions or forces, nor becoming or creation, now, nor projective realisation at the reflective interface of the Projective Realm of the Critical Boundary of the M-Set.

Self-reflectivity, like self-referentiality, is a symmetry of enormous subtlety about which we shall be saying a great deal more as we proceed on our journey in the vehicle of the M-Set, while it is now evident that this symmetry underpins change.

There is no thing without change.

Every thing is change, now, in all ways, always.

7.24 – The M-Set is change, now.

8: About the Force of Gravity

Before we continue to unveil more details about the virtual Quantum Field of the M-Set in order to reveal the forces implicit to the virtual Elemental State of the M-Set, we shall gather together here a number of significant properties of the M-Set's 'ghost force' of the Trilogy of 'intending' that we have now identified as the force of gravity in the Projective Realm of the M-Set which we call the M-Set Universe.

Gravity is the physically realised, albeit illusory force system relating to the Trilogy of 'intending' in the realm of the computational dynamics of the M-Set, the latter being in turn the self-referential influence acting upon the reflective fluctuations of the Tower of Criticalities of the Primordial Quantum Field of the Elemental State of the M-Set.

Gravity is a 'ghost force' because it resides in the Virtual Computational Realm of the M-Set and is associated with the M-Set's 'intentions' for itself in projection.

Gravity arises naturally from the defining property of Self-referentially of the M-Set and will be shown to be fully revealed in the physical sense through the apparent and thence spontaneous breaking of the overriding symmetry of self-referentiality when the Holy Ghost of the Trinity is revealed through the 'Divinity or Division of Unity' of the M-Set in projection with the simultaneous spontaneous breaking of the symmetry of self-reflectivity.

Gravity will emerge in the sections ahead as a purely Quantum Field force of immense subtlety and sophistication that forms the Hilbert Space of States of the fully fledged or wholly renormalised Quantum Field of projective realization that the M-Set is, wondrously and miraculously, as alluded to in the previous section.

Gravity will be seen to be the force system which, in the parlance of theoretical physics, exactly renormalises the Quantum Electromagnetic Field Theory unto 'on mass shell' projection in space and time.

Gravity appears to have a geometric relationship to space and time in the projection but this merely completes the illusion. Space, time and matter are apparitions of the illusion of spontaneous creation, now, which will become clearer too as the M-Set unfurls its secrets.

Gravity as described through the relativistic dualism of space-time is just that, a 'projective description' of a vastly more subtle actuality.

Gravity relates to the Trilogy of 'intending' over all orders and levels of the Tower of Criticalities of the Primordial Quantum Field of the M-Set and, being self-referentially bound as well, the cosmic level universal signal of the presence of gravity will be a closed distribution of signals of Quantum Fluctuations, as indeed has been observed with the so called Cosmic Black Body Background Radiation of the Universe, which has a 'black box' or self-referential distribution.

Gravity manifests its quantum foundations in the M-Set Universe by the 'Cosmic Black Body Background Radiation' distribution, and this is analogous, if you will, to the 'telephone exchange' of TOME (the Tower of Mass Effects) in space-time with the relative scales relating to the orders of sub-tendings implicit to the M-Set. It also follows that gravity and thermodynamical theories of conventional physics are intimately linked in the M-Set Universe, because gravity of the M-Set is precisely that 'ghost force' manifesting in the M-Set Universe as a whole which exactly neutralises dissipation to uphold a perfect critical dynamical equilibrium simultaneously across all orders of mass effects in space and time. Thence too gravity is precisely that apparent physical force system which upholds the 'illusion of thermodynamics'.

Gravity of the M-Set Universe abides the properties of the M-Set and hence, because the M-Set has no nett values, all apparently extant values, qualities or quantities of the M-Set Universe as a whole must have 'duals' unto Oneness at the Triadic Causality of the Critical Boundary of the M-Set which is formed, in turn, by gravity itself. We shall return to speak to the issues of duality in the sections ahead because of its monumental importance in the Projective Realm, suffice it to say here that the M-Set Universe as a whole cannot have a distribution property or structure that is not exactly dual to a dispersion property or function, otherwise the 'neutrality' of the M-Set would be violated. For example, there is no such thing as a Cosmic temperature or 'measure of dispersion' without an exactly dual measure of structure or distributive property while, moreover, all universal dually related properties of the M-Set Universe directly relate to the Gravity Force in projection. The cosmic background 'Black Box' radiation spectrum is, dually, a reflection of both the distributive and dispersive properties of the M-Set Universe as a whole, now, or in this projective cross-section of the M-Set, now.

Gravity as the 'ghost force' of the Trilogy of 'intending' of the Virtual Computational Realm of the M-Set implies directly, as we have seen, that every 'thing' is 'intended' in the M-Set Universe, the inestimable significance of which will become manifestly extended in Part II of this work, while at the most primitive levels of projection it is evident now that statistics and probability theories are mathematical contrivances rather than actualities, reflecting, ultimately, perspectives or points of view as I shall explain in more detail later, rather than 'universal truths' because for the M-Set Universe there are only possibilities. This will also be shown ahead to amount to taking 'projective" cuts' of the M-Set through the simultaneous breaking of the symmetries of self-referentially and self-reflectivity because the M-Set encompasses all possibilities simultaneously, now.

Gravity is a wholly non-singular force system because no singularities can arise through the projections of the M-Set, which itself is wholly symmetric, and therefore the predictions and the objects arising from singularity afflicted theories of gravity have no actuality, nor indeed physical reality, while the intricate cosmic distributive phenomena of galaxies attributed to missing mass on cosmic scales are automatically accounted for by the critical dynamical equilibrium of the M-Set Universe. 'Black holes' do not, need not, and cannot exist in the M-Set Universe. Black holes are figments of perspective taking by observers abiding singularity riddled theories beholden to mathematical formalisms that beget them. As I shall reveal ahead it is possible for the theorists to uphold what are apparently consistent contradictions to the M-Set, the point here being that all the M-Set realisations are ultimately 'observer-other' 'projective cuts' perfectly reflecting the observer's bias, for example, until they choose to change their perspective or point of view.

Gravity is a non-linear, non-singular, dually distributive and dispersive force of immense subtlety in its physical manifestations, as for example in extended matter systems wherein quite generally the gravity of the M-Set

Universe does not have any singular origins ($1/r^2$ Law) but is non-locally originating. Thence too, the 'classical' computations giving rise to black hole collapse of dying stars are misconceived because the dual structural-functional or distributive-dispersive properly of gravity in such collective extended systems is non-localised so that condensations can only occur in turbulent structures' spanning many orders of subtendings, as dictated by chaos and ultimately self-referentiality itself.

Gravity, like all phenomena and force systems of the M-Set Universe, will be seen to abide the 'holographic principle' that the 'information' (informational) content of structure and function relating thereto co-responds in physical terms to the properties of two-dimensional surfaces, while in the actuality of the virtual Computational Realm of the Quantum Computer that the M-Set is, as we shall see ahead, the informational processing corresponds to a manifold (many folded surface) that the boundary of the M-Set which the M-Set is, is.

Gravity is the natural force system of pure turbulence because the 'ghost force' of the Trilogy of 'intending' is the influence of symmetry of self-referentiality itself, unto the critical boundary of the M-Set that the projecting M-Set is, and resulting in turn in the remarkable discovery that gravity is the force system of pure self-organising critical systems in the Physical Universe which we shall come to see as being the M-Set Universe.

Predictably then, if one looks to the signs and signals of chaos in the Physical Universe, the systems exhibiting the purest 'fracticality' are all predominantly naturally gravitating systems such as avalanches, sand dunes, rock faces, coast lines, weather patterns, growth and flow under gravity, cosmic hydrogen clouds, galactic matter distributions, flicker frequency of stars, and on and on.

The apparent cut-offs to pure fracticality and critically arise from the influences of the other force systems, most notably electromagnetism, which we shall discover is not only the force system of the symmetry of self-reflectively itself, but that projective realisation at the Critical Boundary involves the simultaneous apparent or spontaneous breaking of the symmetries of self-referentiality and self-reflectivity while, moreover, it is the apparent simultaneous breaking of both these symmetries which generates the illusory mass, space and time effects at the edge of chaos.

Gravity is the 'ghost force' system of the M-Set which perfectly and non-singularly re-normalises or fashions the Primordial Quantum Field of the Elemental State of the M-Set in every possible way to create, through the yet to be revealed mechanism of spontaneous breaking of self referentiality, the Tower of Mass Effects in space and time.

Gravity is the final arbiter of space, time and matter in the 'physical' universe of observations by light-like signals, this being the relativistic dualistic realm of projection, while the equivalence of gravitational action with classical geometric notions of space and time reflects the deep relationships that reside within the Triumvirate of the Triadic Causality of the projective realisation between the M-Set computational dynamic and relativistic dualisation.

Gravity is automatically fully unified through the 'Oneness' of the M-Set with all the other fundamental force systems residing in the M-Set that will be revealed soon, and all of which are spontaneously created as a result of the defining property of the M-Set, namely self-referentiality.

Gravity is the physical force system that spontaneously appears with the apparent or so called spontaneous breaking of the overriding symmetry of self-referentiality.

Gravity, therefore, like the other physical force systems to be explored later, is an 'illusion of projection' because the M-Set in which this perfect illusion is wrapped up or 'completed' is perfectly symmetric and perfectly free.

Gravity as a force system is responsible for the mass effects of TOME or simply 'mass'. However, in the illusory projection there is no thing; no actual physical object and all mass-effects will be shown ahead, after our exploration of the Quantum Realm, to be self-interactive (self-subtending) self-measurements of the M-Set which creates the illusion of massive objectivity, given 'weight', of course, by 'gravity' itself. These subtleties of the M-Set will be progressively clarified as we gather more experience with the M-Set in forthcoming sections.

Gravity is also the force system of 'complexification' or the manifolding at the Critical Boundary of the M-Set upon the basis of the Primordial Quantum Field of the M-Set. Simply stated, gravity is the force of com-plexity in the M-Set Universe.

Gravity and complexity are one and the same in the M-Set perspective.

Gravity is also the force of renormalisation in the M-Set Universe where the term 'renormalisation' in the parlance of theoretical physics relates to the 'complexifications' of field theoretic processes unto 'on-mass-shell physical projection', while in the M-Set Approach Renormalisation will come to be seen as the action of the Trilogy of 'intending' unto the Critical Boundary of the M-Set of the Projective Realm that the M-Set is.

Gravity and renormalisation are one and the same in the M-Set Perspective.

Gravity is a Dual Force System because its function in projection is exactly dual to be structure it begets although this becomes very com-plexified with the mixing of all of the orders of sub-tendings by the M-Set computations, and thence too the physical force of gravity is extremely subtle and is exactly dual to its physical source, namely the mass effects, in an evidently very complex way. Only in the artificial limit when mass is regarded as a mathematical point that banishes all the extenuated subtleties of the duality of gravity does one return to the singular view of classical theories of gravity. As always, the lore of classical analysis completely denies the actuality.

Gravity is 'the ghost force' of 'intending' of the Trilogy of the M-Set computations and, being related to all orders of reflective fluctuations in the M-Set Quantum Field, gravity is a scale invariant force system in projection meaning that the nature of the force is the same, regardless of scale, in the physical sense, while such a scaling is also not naïvely broken by mass because gravity 'sees through' mass down to the Quantum Field level of the M-Set. In other words, mass is completely transparent to gravity, just as mass itself is an illusion of the Dual Force System of gravity in the projective realm. Scaling only appears to be broken or is spontaneously broken by the dual nature of the Gravity Force System because

8.1 – the M-Set has no scale.

Gravity as a force system simultaneously and spontaneously or 'apparently' breaks the no scale symmetry of the M-Set. Stated alternatively, you can not actually break scaling symmetry if there is no scale in the first place. As we shall again explore soon, the unimaginable complexifying power of the mixing of the self-referentially begotten Pure Chaos of the M-Set computations creates the con-fusion or apparition of apparently scale-breaking 'mass effects', and as always, and in all ways,

8.2 – The M-Set is the master of illusion.

Gravity as a Dual Force System in projection is con-fusion caused by the Pure Chaos of the M-Set. Even our language abides actuality!

Gravity ultimately is a force system deriving from information exchange or mutual information implicit in the reflective fluctuations of the virtual computational support that the M-Set is, and, as a consequence too of the purely chaotic nature of the computational dynamic of the M-Set, associated as it is with maximal mixing, maximal mutual informational exchange, and self-referential, self-'intending' criticality, it follows that all gravitationally influenced phenomena will exhibit signs and signals or footprints of chaos. Indeed, this is most evidently the case on cosmological scales with fractal distributions of matter, fractal flicker frequency distributions of luminous bodies, and the attractor characteristics of orbits in extended perturbed systems all the way up to the perfectly balanced clusters of stars in galaxies as well as the coherent shapes of cosmic dust clouds even though they span hundreds of thousands of light years.

Gravity, quite literally, is an information exchange force system from the perspective of the virtual Computational Realm of the M-Set and which force system is present at every level of systemic complexification in the M-Set Universe, thence in all generality and for all possible systems the processing of information by the M-Set computations gives weight to the systems which refers back, in turn, to the fact that the processing of information by the M-Set spontaneously creates the structural conduit for the intended function dual to do it, this being the Realm of Dual Relativistic Projection of the M-Set.

Gravity is 'quantised' by the 'virtual exchanges of information' in the computational support that the M-Set is, which we shall unravel further in the section on Quantum Mechanics, while the nature of this is the same over all scales and corresponds to the reflective fluctuations unto the Critical Boundary of the M-Set where Triadic projection occurs.

Gravity, topologically speaking, is like an unimaginably complicated (multiply folded) virtual, reflectively fluctuating or vibrating membrane in the indefinitely dimensioned Hilbert Space of the fully renormalised Quantum Field that the M-Set is. This picture will emerge more clearly from the M-Set's foundational relationship to Quantum Mechanics in the sections ahead.

Gravity as a physical force has no absolute strength because the M-Set has no absolute measures and therefore the physical strength of gravity is 'relatively' determined and will be shown to ultimately relate back to the 'quanta' of Gravity and to the perspective or 'projective cut' when a fuller realisation of the M-Set is established ahead.

Gravity as a Quantum Field force derives its quantum basis directly from the 'self-referential' actions of the virtual 'ghost force' of the Trilogy of 'intending' upon the primordial Quantum Field of the Elemental State, a phenomenon known in the parlance of physics as 'renormalisation' while it will become evident in the sections ahead when we explore the M-Set's role as the founding stone of Quantum Mechanics that the origin of the Quantum action of gravity resides with or is implicit to the reflective fluctuations or 'primordial quantum actions' of the elemental state itself that generates the Primordial Quantum Field and which, in turn, the Gravity Force System elaborates into the Critical Manifold by the Trilogy of 'intending'. Evidently then, the mani-folding 'actions' of the Gravity Force system in the Virlual Realm of the M-Set will themselves be very 'complexified' unto the Critical Boundary of the Projective Realm, the point we wish to emphasise here being that the nature of the gravity Force System in projecton is a Critical Manifold with unimaginably complexly ramified quantum actions, while acknowledging that the origin of the 'quantisation' of the Gravity Force System is with the primordial computations of the M-Set that we shall continue to explore further in the sections ahead.

Gravity from the perspective of the Projective Realm acts topologically like a manifolding force of the Primordial Quantum Field of the elemental state of the M-Set, while all projection, as we shall see ahead, is a 'light-like' self-differentiation of the M-Set from the level of the 'Sheet' or 'Primordial Quantum Field' that the Gravity Force System manifolds or multiply folds on itself. Thence, in effect or in the 'projection' of the action of the Quantum Gravity Force System of the ghost force of the Triology of 'intending', the relative strength of the physical force of gravity in comparison to, say, the Electromagnetic Force System (that we shall discover ahead derives its 'quantum action' directly from the elemental state) will relate to the compounded or 'complexified actions' of gravity *vis à vis* the singular quantum actions of electromagnetism. Drawing on a 'physical analogy' here, this is like comparing the strengths of the bonds of molecules in the sheet of a piece of paper to the strengths of the bonds between the folds of the sheet. Carrying this analogy into the context of the M-Set, the now 'self-referential sheet' self-folds on itself while simultaneously it is the property of self-referentially itself which spontaneously conceives the elements of the sheet in order for the sheet to self-referentially self-manifold. This is also the miracle of the M-Set, as we shall continue to discover ahead, which is all the more profound when we come to appreciate that gravity is the force system of the 'self-referential' 'ghost-force' of the trilogy of 'intending' of the M-Set that manifolds the Primordial Quantum Field of the Elemental State in precisely the 'critical' (self-referential) way required to effect the apparent or spontaneous breaking of the symmetry of self-referentiality begetting the same in order, in turn, for the Projective Realisation of the M-Set to occur.

Consequently we shall also come to appreciate the monumental subtleties of gravity which, because it is the force of the symmetry of the defining Property of the M-Set, will ultimately be seen as the Holy Ghost of the Holy Grail of One that the M-Set is that unifies all phenomenology of the Projective Relam.

Gravity is the force of One and, as we shall discover ahead, there is ultimately only One force, the Gravity Force System of the 'ghost force' of the trilogy of 'intending' that begets and unifies all other physical forces of the M-Set Universe. This is the Magic of the M-Set.

Gravity is a dual force system in projection, as indeed all force systems of the M-Set projections will be shown to be because, at the Triadic projection, the M-Set is critically balanced (self-referentially) as a 'from within' dualisation of structure into the space of function of the 'redundant' or 'reduced out redundant information' begetting the same (or read, for example, 'space of freedom of invariance'), the point here being that the Gravity Force System involves the duality of structure and function manifesting as a perfect dynamical equilibrium over all scales of informed structure and subtending functions. The Gravity Force System is thence simultaneously the attraction into structure and the dispersion into function which, from the defining property of self-referentiality of the M-Set, are perfectly dynamically and critically balanced across all levels of the M-Set Universe.

8.3 – The M-Set Universe as a whole is a perfectly dynamically balanced critical state, in all ways, always, now.

Because of the mixing of the levels of the Trilogy of 'intending' of the fully renormalised M-Set Quantum Field by the Gravity Force System, and also as a result of the 'points of view' of local perspective taking, the holistic purity of the property 8.3 is distorted in the illusion of the projection to create our observations and experiences of it. This aspect of projection and variants of it is also highly significant in comprehending how 'local view points' or 'perspectives' of the M-Set are created in projection and will be discussed further in the next section.

Gravity as a 'dual force system' maximally dualises into structure and function by virtue of the M-Set properties. However, as a result of the informational exchange nature of gravity, what is one level's intending' is 'another level's extending or sub-tending', and so it is that ultimately all information exchange 'weighs', while it will depend in the end upon one's perspective, level of observation, or 'point of view' as to what the 'mass effect' is in the Projective Realm. The nuances of gravity are subtle in deed.

Gravity is a wholly non-local, scale invariant, non-linear, critical, dual force system at the level of the M-Set, and it is the nature of the 'light' projection at the Critical Boundary of the M-Set that creates the 'local' relativistic apparitions of the force system, including the apparitions of Einstein's equivalence principle of the Theory of General Relativity, while the M-Set Universe as a whole is One, now. We shall have more to say about this when exploring the Quantum Mechanical nature of the M-Set and the phenomenon of 'apparent' localisation or 'breaking of non-locality' by the all important property of the apparent 'collapse of 'wave functions' or the 'representations' while we remind ourselves here that all such phenomena and properties are underpinned by the overriding actuality that the M-Set is all that is, now.

Gravity can only be comprehended as a phenomenon of the M-Set of which its actual quantum properties are also borne while, as we have seen now, the unimaginable 'complexity' of the con-fusion of the projective realm, the nature of which depends upon one's prospective or point of view, could never be unravelled by hand mathematically to reveal the actual origin of gravity, which the M-Set is. The secrets of gravity reside with the Defining Property and overriding symmetry of the M-Set, namely self-referentiality.

Gravity is automatically a 'super symmetric' force in the sense that this term is used by physicists because the 'overriding symmetry' of self-referentially includes this symmetry of physics, which links the so called 'bosonic' and 'fermionic' degrees of freedom. This symmetry will be seen ahead to follow from the perfect duality of structure and function of the M-Set Projective Realm that reflects back to the fundamental implicit operand/operator duality of the M-Set and which, when referencing this all back, in turn, to the foundations of Quantum Mechanics, will link the 'bosonic' and 'fermionic' degrees of freedom to the functional and structural aspects of the M-Set projection respectively.

And so it will be borne out too in the sections ahead that

8.4 – the M-Set only projects dual force systems, while it is also the case that

8.5 – the M-Set spontaneously breaks its own symmetries to generate the illusion of force systems.

And, moreover,

8.6 – the M-Set Universe force systems effect functions which are exactly dual to the structures sourcing them with the major qualification here that the 'sources' are unimaginably intricately interwoven dualities of structure and function, the complexity of which belies the perfect symmetries underlying the same, such as the symmetry of self-reflectivity which the mixing dynamic of the self-referential computational property of gravity so completely interweaves that it creates the illusions of space, time and matter with the apparent breaking of this symmetry.

We now know that

8.7 – the M-Set self-renormalises unto the criticality of self-projective self-realisation, in all ways, always, now.

Miraculously, gravity is the physical force subserving this property as well as the property that

8.8 – the M-Set Universe is a perfect dynamical equilibrium on the edge of chaos as demanded by the defining property of Self-referentialty that is 'the vessel' of pure chaos.

The M-Set Universe is in perfect harmony and balance.

The current General Theory of Gravity is a mathematical model of a singular view of gravity founded in relativistic dualism.

Gravity is not that.

Look again to the Heavens.

What is your re-vision telling you now?

9: On the Nature of the M-Set Projective Realm

The M-Set Projective Realm, as we have already alluded to on a number of occasions, is a truly Alice in Wonderland world of illusionist magic shimmering at the 'Critical Boundary' of the M-Set which, from the primordial property.

9.1 – the M-Set is the boundary of itself, implies that the M-Set Projective Realm is the M-Set. And so it follows directly that

9.2 – the M-Set is the projection of itself which is generated, as we shall continue to discover, by the process of self-referential, self-reflective, self-representational, self-realisation, or SR4 for short.

Because of the property 9.2 the projective realm is automatically provided with all the properties of the M-Set. However, this is cleverly and complexly camouflaged by, for example, relativistic dualism and relativistic causality as well as the mixing of the critical orders of the Primordial Quantum Field by the Gravity Force System to create the TOME (Tower of Mass Effects).

From our discussions earlier and the properties of the M-Set revealed thus far it is apparent also that

9.3 – the M-Set is all possible projections of itself in parallel. Therefore, the so called 'deep well of possibilities' of the M-Set is a self-referential Triumvirate consisting of the Duality of the Projective Realm and the 'Oneness' of the M-Set.

It is not without cause then that we regard 'reality' or projective realisation as dualistically relativistic. However, it is now clear that First Cause arises from the Trinity of the 'Oneness' of the M-Set and that without the Triadic Casaulity relativistic dualistic causality has no basis for its 'existence'.

The 'spontaneous' or apparent breaking of symmetries of the M-Set about which we shall have more to say soon leads directly to the observation that

9.4 – the M-Set is the cause of spontaneous creation, in all ways, always, now, and thence

9.5 – the M-Set is the beginning, now
Because the beginning is the projection 'now' which is borne on light-like representations it is also the case in the M-Set perspective that in the beginning there is light. And no thing beside this.

Now, as we mentioned earlier, each M-Set projection is like a division or a cut, or even a cross-section of 'the deep well of possibilities', and which phenomenon is also dependent upon the apparent or spontaneous breaking of the symmetry or reflectivity the mechanism of which will be revealed in detail in the section on the Quantum Computer properties of the M-Set. Suffice it to say here that the symmetry of reflectivity has profound implications for the Projective Realm which we pre-empt with the property that.

9.6 – the M-Set projections reflect its Divinity. In other words, at projection the 'division of unity' or 'Divinity' is a reflection of the M-Set that exactly ensures self-referential closure.

Moreover, we observe now that the 'Divinity' of the M-Set is the self enlightenment of the self realisation of the M-Set.

Let us reflect here for a moment on the self-reflective symmetry of the M-Set and the realisation that Divinity and self-reflection are intimately linked in the Projective Realm of the M-Set.

Clearly, and with reference to our earlier discussions, the self-reflectivity of self-referentiality is expressing in the reflective fluctuations between representations of the Primordial Quantum Field of the M-Set which, in turn, is closed self-referentially in the M-Set, while simultaneously the 'influence' of the Trilogy of 'intending' of self-referentiality itself or the ghost force of gravity is mixing the Tower of Criticalities of the reflective fluctuations of the Primordial Quantum Field into very complex structures and functions. Pure one to one reflections of say, an unmixed state of the Primordial Quantum Field of the modes of the virtual Element State of the M-Set become extraordinarily complicated (folded into each other over all orders, over and over), the point here being that the M-Set is now acting self-dually like a Matrix operator (abeit, unitary) which mixes fluctuations of itself in all possible ways (recalling that 'operator' and 'operand' are exactly dual in the self-referential M-Set).

Let us recall here that because of the symmetry of self-reflectivity, any projective cut of the M-Set 'now' is an exact reflection of itself under the umbrella of self-referentially. However, in reality, because of the 'mixing' dynamic of the M-Set this symmetry is apparently spontaneously broken, which we shall learn more about in the section on the Quantum Computer properties of the M-Set. If we focus here just on microscopic or very elemental reflected aspects of the M-Set that are relatively unmixed, such as charge, parity, or the handedness of neutrinos and DNA, we see that they are characterised by elemental but apparently 'universal' 'asymmetries', the asymmetry in these cases being resolved, as we shall learn later, by the presence not only of an implicit reflected 'dual' but by the criticality of the Projective Realm to ensure closure self-referentially.

With regard to phenomena relating to the 'elemental representative' states of the M-Set the 'connecting' reflective fluctuations (ghosts of the M-Set) will turn out in projection to be the 'force fields' mediated by the 'quantal actions' or reflective fluctuations of the M-Set computations.

Thence, the so-called force field around a charge, for example, is because this charge is, in effect, 'looking at its reflection' or image by way of the 'reflective fluctuations' or quanta of the M-Set, while for the ghost force of gravity reflective fluctuations over a vast range or manifold of levels reflecting the informational complexity of the particular system are at play, as alluded to already in the enumeration of the properties of the M-Set Gravity Force System and which will be elaborated upon soon in subsequent sections.

We are anticipating here aspects of the general nature of force systems manifesting in the Reflective Divinity of the projections of the M-Set, including those of the weak, strong and electromagnetic force systems implicit to the elemental representative states that are 'renormalised' and 'universalised' by the proliferative amplifications of the M-Set computations which equate generally, as we now know, to the action of the ghost force of gravity. Remarkably, it will be seen to be the case that all force systems are 'ghosts' of the reflective fluctuations of the M-Set which manifest in the projection as precisely those quantised force fields needed to uphold the self-referentiality of the M-Set, and this is also precisely the circumstance at the Critical Boundary of the M-Set where the criticality of the M-Set computations projects simultaneously with the apparent or spontaneous breaking of symmetry.

The highly interrelated subtleties of the M-Set are no more intertwined than in the apparent emergence of quantised force systems in the Reflective Divinity of projection.

The Divinity or 'Division of Unity' of the M-Set that occurs in relation to the reflective fluctuations and which involves the process of self-referential. self-reflective, self-representational self-realisation corresponds to a self-differentiated projective cut or cross-section of the M-Set, and we shall continue to explore the properties of the Projective Realm here, beginning with the observation that

9.7 – the M-Set projective cuts are concomitant with the appearance of dual force fields while it is also the case that

9.8 – the M-Set projections are concomitant with spontaneous symmetry breaking and thence, as we shall learn later,

9.9 – the M-Set projective dual forces systems are precisely those forces systems which ensure the exact preservation of the M-Set symmetries begetting them, which then leads to the very general property that

9.10 – the M-Set projective dual forces systems are precisely those forces systems that exactly uphold the illusion of the spontaneous symmetry breaking which begets them, which underlines, in turn, the complete illusory nature of force systems in the projective Realm of the M-Set.

It is also the case that

9.11 – the M-Set projective cuts simultaneously involve all levels of the M-Set manifold.

So, in effect, each projective frame reveals the complexity (manifolding) of the mixing of the self-referential, virtual, computational dynamic of the M-Set. This property in turn leads to a highly significant and subtle property of the Projective Realm which we shall have occasion to explore again, namely,

9.12 – the M-Set projections are a mixture of all their histories. A naïve exemplary statement of this is that it would not be possible to have a projection involving the planet earth without the historical context of the heavens wherein is written the birth and death of it all, now, from the hydrogen gas clouds to star nurseries, and on to the elaboration of the chemical elements, planetary formations, and to star death and re-volution. This will also become clearer when we explore the Memory properties of the M-Set.

9.13 – The M-Set projections are all in context,

rather like the phenomenon of our 'identity now' being the sum of our histories.

From the property that

9.14 – the M-Set projections are at the Critical Boundary of the M-Set,
we shall have occasion soon to also focus our attention on the Critical Boundary where the Triumvirate of the Triadic Causality resides, embodying as it does both the Oneness of the M-Set and the Duality of the projective Realm, while it follows too, of course, that because the M-Set is the boundary of itself

9.15 – the M-Set is simultaneously the projection of itself, now, always, always.

The Critical Boundary or 'criticality' of the projecting M-Set also implies, in consonance with the 'light-like' representations of the M-Set at the Critical Boundary, that the M-Set projection generates the illusion of a three-dimensional physical world through change at the Critical Boundary which, in physical terms, is like a statement of an holographic principle of three-dimensional projection that will become more evident when we further explore the Quantum nature of the M-Set in relation to a deeper exploration of the Critical Boundary in the sections ahead. Suffice it to say here that the Critical Boundary can be likened to a spontaneously fluctuating manifold of convoluted re-flective fluctuations of unimaginable complexity involving the mixing or 'folding-in' of the Primordial Reflective Fluctuations of the Elemental State (the representation of which we previously referred to as a Tower of Criticalities) into ever more complicated critical levels unto the Tower of Mass Effects of the fully renormalised or fashioned Quantum Field of the M-Set under the influence of the Trilogy of 'intending' of the ghost force of gravity.

As we shall see in the sections ahead, the Primordial Quantum Field of the Elemental State of the M-Set is the substrate of the 'subatomic' world of forces which, if you will, is the 'thread' with which the ghost gravity force system weaves the fabric of the Cosmos. The layers upon convoluted or interwoven layers of the renormalisations or fashionings by the Gravity Force System create, in effect, a complexification of the primordial

Quantum thread to generate the Critical Tower of Mass Effects at the Critical Boundary of the M-Set where the symmetries of self-referentially and self-reflectively appear to be simultaneously and spontaneously broken. But they are not. This is the miracle of self-realisation of that which is One.

Implicit to the projection then is the Subatomic Quantum Realm and its reflective dual, implicit to which in turn are subatomic force systems of the M-Set that will be revealed in the following sections and which, as we shall also come to see, are woven by the gravity force system into their physically manifesting fashioned or dressed, or 'fully renormalised' forms in the context (fabric, texture) of it all, now.

Evidently too, if we are looking at projections for which the focus of observation or measurement is at the level of subatomic phenomena and where complexity is also minimal then we shall see fundamental asymmetries such as parity, charge, time direction and handedness, relating as they do the deeply embedded symmetry of self-reflectivity of the M-Set, whose role in the scheme of things will be seen to be prominent when we turn our attention to the Quantum Realm and the Quantum Computer that the M-Set is. Suffice it to state clearly here that at the level of observations of elementary particle physics the reflective duality implicit to the M-Set is apparent.

The beautiful actuality is that the apparent breaking of the symmetry of self-referentially of the M-Set is accompanied by elemental or subatomic force systems, the apparition or appearance or realisation of which, together with the apparent or spontaneous breaking of the symmetry of self-reflectivity itself, ensures exactly the preservation of self-referentially while these apparent effects are simultaneously locked-in by the 'mixing dynamic' of the M-Set which, in turn, renormalises itself in all possible ways unto the Criticality of the boundary that it 'is' where the symmetry of self-referentiality is apparently broken. Simultaneously, the spontaneous breaking self-referentiality ushers in the Gravity Force System, which is exactly that apparent force to ensure the preservation of the symmetry of self-referentiality in relation to the apparent Tower of Mass Effects (space, time, matter).

Evidently

9.16 – the M-Set projections are 'apparent', and because

9.17 – the M-Set projections are exactly symmetrical unto self-referentiality

it is the case too that

9.18 – the M-Set projections are dual apparitions unto the perfect symmetry of the M-Set.

Also, because

9.19 – the M-Set projective properties and principles are those of the the M-Set at Triadic Casualty it follows that

9.20 – the M-Set universe is perfectly symmetrical, in all ways, always, now.

Hence to every effect in the M-Set Universe there is an exactly balancing counter-effect unto the perfect symmetry of it all.

For measurements of observations of the Projection Realm at more complex levels a new and startlingly significant apparent asymmetry becomes evident, namely the apparent asymmetry between the object and the context, the self and the other, the 'this' and the 'that', the chicken and the egg, the organism and the environment, the observer and the observed, endlessly. But this asymmetry is an illusion too because the M-Set Universe is One, and so it follows that, regardless from which order, aspect, perspective, point of view, systemic closure, or level of com-plexity the projection is observed, for the M-Set Universe as a whole,

9.21 – the M-Set projection of 'that' is wholly 'reflected' in the 'other'.

In other words, while projection creates the illusion of separateness no thing is separate in the M-Set Universe and every aspect of the M-Set universe is wholly reflected in the other. It is all One.

9.22 – The M-Set is thoroughly post-modern.

From the illusion of separateness and the property of Oneness of the M-Set, together with the property about which we shall reveal more later that the M-Set self-differentiates through the Trinity of Oneness unto self-realisation, there follows a profound property whose ramifications will resound in our future discussions, namely,

9.23 – the M-Set resolves all dualities by the Trinity of the Oneness that it is.

The 'illusion of separateness' and apparently unresolvable relativistic duality are one and the same in the Projective Realm of the M-Set.

In order to illustrate some aspects of the properties of the Projective Realm of the M-Set to this point we shall consider here the example of a fossil bone of which one asks 'what is the age of this bone?.' In truth this bone has no age because it exists (ex-is in time and space). When one looks at the 'projective cross-section represented by the bone' it is readily appreciated that it is a simultaneous mixture of histories involving the fusion of atomic elements in stars, the elaboration of molecules in biospheres, the assemblage of DNA in the evolutionary story from primitive life forms, the life cycle from birth to death of the organism, the geological story of the excavation site, all now. The bone is indeed an intricate mixing of nested universal histories, now, like the cake before you is a mixture of its ingredients, now.

The fossil bone is a projection of itself reflected in the other, now.

All things in the M-Set projection we call the M-Set Universe are the sum of their histories.

9.24 – The M-Set is the mix-master of all possible histories.

Neither the fossil bone nor the cake are truly separate from their context. Every level or scale of the sum of histories of the fossil bone is reflected in the other, now.

We shall see in discussions ahead how the causality of projection is, in truth, the 'vertical' parallelisms of histories of the M-Set rather than the apparent horizontal or linear causality of relativistic dualism.

Duality is an illusion of the projective realm of the M-Set. It apparently exists only, and like relativistic causality dualism is not part of First Cause. It is only an effect, albeit an illusory effect of the Trinity of the Oneness that the M-Set is.

It will become clearer too as we proceed that

9.25 – the M-Set Projective Realm is cyclical, in all ways, always.

For the M-Set there is no beginning and no end in a linear sense of causality. There is only now, like the point on a circle where the 'this' and the 'other' are joined to complete the whole or fold together as 'One', always.

Hence too

9.26 – the M-Set is all possible projections of itself.

As the ghost of gravity weaves its 'intentions' in the fabric of the Primordial Quantum Field, so too do the apparent reflective asymmetries of the Projective Realm become simultaneously more and more self-referentially focused, influenced or modularised in all possible ways unto the complexity of the object or the individual that is being wholly reflected in its context or environment at the Critical Boundary of the M-Set.

Separateness or 'disunity' is the profoundest illusion of the M-Set Projective Realm and gravity is the diviner of the illusion.

In the Projective Realm of the M-Set the apparent reflective asymmetry between the 'this' and the 'other' is the manifestation of the myriad of 'intentions' of all levels of the complexifying manifolds of the M-Set, from the raw physical mutual attraction of the body to the planet Earth to the nuances of mutual empathic attraction of emotionally reflectively connected interacting 'individuals' states of mind.

The 'intending' forces or forces of the Trilogy of 'intending' are all of the same nature in the M-Set, amounting therein to the information exchanges or 'mutual information exchanges', and thence to the com-putational processing of the information of the reflective fluctuations of all possible 'intentions' of the complexifying renormalisations of the Primordial Quantum Field of the M-Set, all of which give 'weight' to the projection.

The gravity system is immensely subtle, unto all levels of projection of 'complexity'. All information or creation in the Projective Realm is 'intended' or gravitating self-referentially over all levels of complexity of projective realisation.

9.27 – The M-Set projection is spontaneous creation. And

9.28 – the M-Set spontaneous creation is intended.

Deferring also to the complete parallelism of projective possibilities

9.29 – the M-Set projects parallel universes, now, while

9.30 – the M-Set Universe is a parallelism of all possibilities of itself.

To understand the significance of the latter property we need only defer to the llusion of separateness of disunity to realise that every entity or individual is referencing an apparently separate parallel universe, but, because 'the other' wholly reflects 'that', this immense parallelism is wholly unified or 'One' version of itself, the M-Set Universe.

Moreover, the parallelism of 'subtending' universes, such as, for example, individuals interfacing as a group, relates to higher and higher orders of reflective fluctuations and complexified mass effects, yet in all ways and always in a wholly self-consistent 'unified' manner under the umbrella of self-referentiality begetting it all, now, while this example further illustrates the unimaginable subtlety and complexity of the Gravity Force System manifesting in the parallelisms of the projective realm. Needless to say, the 'intended' exploration by the M-Set of its self-referentially generated unified parallelisms of possibilities of itself defies a complete description.

At this junction we return more specifically to the illusions of space, 'motion' and matter in the Projective Realm to advance the elucidation of the properties of this realm. Because the M-Set projection is just that, namely a projection of itself, the only 'actuality' is the M-Set itself, which 'is that' so we therefore need to be able to apprehend in a more tangible or understandable way the basic physical illusions of the Projective Realm all of which relate to the grand illusion of 'separateness' or disunity from which alone one derives the senses of object and 'other' or of 'thing'. And also of this thing 'here' or space and volume, and of 'change' of thing and motion, and thence of time.

Let us focus first on change with the recognition here that without change there is no self-differentiation of the M-Set and no projection or no thing, while evidently

9.31 – the M-Set Universe is 'change' in the context of the Defining Property of the M-Set, namely self-referentiality.

Now, when asking any question of the M-Set, such as the question 'How does the M-Set change?' one must appreciate that as the self-referential M-Set has no absolutes the answer involves simultaneously apprehending a 'triumvirate' of concepts because all fundamental aspects of the M-Set are related in threes or Trilogies 'reflecting' the Trinity of Oneness and the triumvirate of Triadic causality in contrast to our habit of dualising the conceptual understanding of reality based upon the manifest history of dualistic relativism, while, just as an aside, a deeper causal understanding of even the post-modern view of 'object' and 'other' can never be achieved by duality alone because we are always left holding the two ends of the pieces of the chicken-egg string without the connection! Without, in deed, the closure or the completion that the M-Set provides, as we shall return to in Part II.

Without symmetry or 'redundant information' there is no 'room' for change and in the purely self-referential Virtual Realm of the M-Set it is symmetry that provides the 'freedom' for change while we now know

9.32 – the M-Set is maximally symmetric and maximally free; its only constraint being the unlimited symmetry of self-referentiality itself, and thence all change is for free in the M-Set.

Moreover, because of the symmetry of the M-Set, which provides the context of the Defining Property of self-referentiality, it follows that

9.33 – the M-Set changes freely, while it is also the case that

9.34 – the M-Set changes spontaneously, because the M-Set has no time.

It follows directly too that

9.35 – the M-Set projections are for free because

9.36 – the M-Set is free through its 'complete' symmetry. Therefore it is the case that

9.37 – the M-Set Universe is created spontaneously for free and also that

9.38 – the M-Set is freely creative.

All physical illusions of the Projective Realm of the M-Set as a whole are both perfectly Dual and identically One at the Triumvirate of the Triadic Causality of the projecting M-Set, even unto the Illusion of Separateness of the observer and the other.

There is no such quantity, quality or value as zero in the M-Set Approach and in the M-Set Universe because

9.39 – the M-Set has no 'zero' origin of itself.

Rather

9.40 – the M-Set is the origin of the M-Set.

There is, therefore, no absolute vacuum in the M-Set Universe.

The M-Set Universe abhors a vacuum!

There are also no 'givens' other than what is for free in the M-Set Universe and all apparently extant qualities, quantities and values in the M-Set Universe are dually paired identically unto the Oneness that the M-Set is.

Because the M-Set is One and because the qualities, quantities and values of all things of the Projective Realm are determined in projection by dualistically bound reflective self-differentiations we arrive at the profound property that

9.41 – the M-Set is self-measuring.

Alternatively stated,

9.42 – the M-Set projections are self-measurements of the M-Set.

Self-measurements can only occur if the M-Set can self-differentiate, which relates back again to the reflective fluctuations and thence to change made possible by the innate symmetries of the M-Set, notably self-referentiality and self-reflectivity. And so it is that

9.43 – the M-Set symmetries beget the symmetries of the Projective Realm.

'Symmetry' here is another word for self-measurement or 'sym-metry' in the M- Set perspective, and symmetry occurs simultaneously with the spontaneous breaking of the symmetries of self-referentiality and self-reflectivity as we shall appreciate in more detail when we come to reveal the foundations of Quantum Mechanics implicit in the M-Set.

We know now that

9.44 – the M-Set is maximally self-interactively self-measuring from what we have learnt to this point about the Virtual Realm of the M-Set and about which we shall learn more soon from the Quantum computational properties of the M-Set, while it is through these self-computations that

9.45 – the M-Set gives 'change' for free' to the Projective Realm.

The symmetry of self-referentiality freely allows representations and thence reflective fluctuations and change with the inestimably important observation here in the context of the M-Set that all of these phenomena are actual rather than merely being contrived constructs or mathematical symbols on a page because the M-Set alone is dictating the terms and spontaneously generating the same. We are observers of its actuality, and so it is that our observations have led us to the astonishing property that

9.46 – the M-Set is alive or spontaneously and freely changing.

Life is spontaneous and free change.

9.47 – the M-Set is the origin of life.

Moreover,

9.48 – the M-Set manifold Universe is alive, in all ways, and always.

The rock is alive. The rock is changing, now. Without change there is 'no thing'.

In actuality, the M-Set is freely teeming with representations of itself relating to all the freedoms or symmetries contained by the Holy Grail of self-referentiality. But, this 'activity' which is in the realm of the M-Set is virtual until it becomes projected; however, in truth too, the M-Set maintains its virtuosity because all the illusions of physicality uphold self-referentiality exactly through the Divinity of the Self enlightenment, as we are about discovering on our journey.

We note here that we talk of phrase or 'Θ' as the archetypical variable of redundant information and thence of change implicit to the M-Set because this variable itself cannot have any dimension relating to a physical thing, nor dimension of space or time, while we know that representations of the M-Set must all be 'unitary' or normalised to one so that all iterates, for example, self-scale relatively by the 'order' of replication which in M-Set terms is a form of self-division maintaining a 'norm' of one. Thence, when we try to imagine change in the M-Set we shall often defer to the mathematico-physical idea of 'phase change' in line with our earlier discussions of the notion of 'phase' or 'phase space' implicit to physically active systems.

As an aside, we note here that in order to understand the imaginary or Virtual Realm of the M-Set we have to use our imagination and in Part II of this work we shall come to see how our imagination does indeed relate faithfully to the Virtual Realm of the M-Set.

It is evident also from the self-referentiality of the M-Set that

9.49 – the M-Set can only change by self-representation, so we now have the Triumvirate (symmetry, representations, change) underpinning the spontaneous activity of the M-Set which manifests protectively at the Critical Boundary of the M-Set as virtual phasal waves or the'coherent fluctuations' of the Quantum field of the M-Set through the phenomenon of exponentiation of the reflective fluctuations, which phenomenon we shall revisit soon.

Symmetry is also synonymous with redundant information or equivalent, indistinguishable representations, but the redundant information is wholly irreducible. That is, the redundant information cannot be organisationally or structurally condensed to a 'mneumonic' of itself. What we shall come to understand by structure resides with the invariants of symmetry which by definition remain unchanged by the symmetry; however, an invariant or structure only realises through change or, in the M-Set sense, through representations while the 'redundant information' of the M-Set representations is expressing in dummy variables that reside within 'phase' variables like 'Θ' as we shall see soon through the Quantum Mechanical properties of the M-Set, and which are 'expotentiated' or explicated (unfolded) by the phasal waves 'exp i Θ'.

The process of critical projection with 'exponentiated' is also a manifestation of the property that

9.50 – the M-Set maximally reduces out redundant information at projection.

Alternatively stated, the 'exponentiation' of the representations maxially presents the redundant information of the M-Set to the projection whereupon the irreducible redundant information creates the illusion of space in the Projective Realm in which change related to the representations of the M-Set is realised (made to look real).

Furthermore, as we shall explore in more detail in the next section, the 'invariants' of the representations, and thence the invariants of the symmetries of the M-Set that beget the freedoms for the representations of the M-Set, contribute to the structure of things or to the 'quantisation of mass effects' of the projective realisation. However, in truth, the M-Set performs an amazing magical trick.

Because of the simultaneity of both change and of change giving definition to the 'invariant of the change' which is also begotten thereby it is the case that structure and function spontaneously dualise in the space of the redundant information begetting the structure whose changes in that space simultaneously exactly beget it or, dually speaking, simultaneously functionalise it into existence. And so it is that we are able to say 'everything' is change', and, without change everything would vanish.

To put the above statement in simpler language, the change that begets the 'invariants' defining structure is also the function of the structure that is begotten thereby. In other words too, in the M-Set Approach structure functionalises itself into existence in a perfectly dual way because the structure's function simultaneously defines the structure that is functionally expressing, and this pertains simultaneously for all levels of the Projective Realm of the M-Set as a whole, now, which we call the M-Set Universe.

Clearly at complex systemic levels in the M-Set Universe there is a Manifold of mixed critical levels when structure and function will become very entwined, as we shall discuss ahead, and exemplified by the neurotransmitters in the brain being both molecular structures and 'functional vectors'. The point we wish to draw attention to here is that by taking local perspectives of the 'projective cuts' of the M-Set the pure dualism of structure and function of the M-Set Universe as a whole will appear to be partial or broken, while, on the other hand, for very specific and isolated observations such as those of highly selective elementary particle physics experiments wherein mixing of levels of complexity is also minimal the symmetry of duality as it pertains to 'structure and function' or to matter and forces may appear to be pure. But it is only approximately pure because the parameters of the observation process itself spontaneously breaks symmetry as the discussion here illustrates, and as will become much clearer when we understand the Quantum Mechanical nature of the M-Set ahead.

Indeed, the M-Set simultaneously represents itself over and over, freely, now, always, in all ways, and it is the representing of itself or, if you will, the freely symmetrical self-differentiating self-transformations of the M-Set that create the illusions of matter, space, motion and, consequently, time, all of which are illusions of the Projective Realm.

The M-Set acts like a unitary, albeit unimaginably large universal Matrix of its own self-differentiating self-representational self-transformations, and this Matrix is like the engine that drives the M-Set through the maze of its virtual realm of possibilities, or like the modem of a computer steering a path through the pre-programmed space of memory of a CD-ROM game.

Perhaps one of the most subtle of the illusions of the projection from a mathematical perspective, and which we shall also discuss further when we elucidate the M-Set's role as the foundation of mathematics, is the illusion of a continuum. Suffice it to say here that the 'continuum illusion' is directly related to the M-Set's reducing out the irreducible 'redundant information' of its implicit symmetries or, phrased in another way, the illusion of a mathematical continuum in the realm of Projective Realisation is created by both symmetry and the 'irreducibility' of the redundant information corresponding to that symmetry.

Finally, in this section we acknowledge that

9.51 – the M-Set is constant change, now, in all ways, always, which alludes to the great paradox of change we shall have occasion to discuss in other contexts, while it is timely to note here that the notion of 'constant change' is also a deeply Quantum mechanical and a Quantum field phenomenon which we hold up before us as we enter the next section.

10: The M-Set and the Foundations of Quantum Mechanics

On a number of occasions to this point attention has been drawn to the naturally arising Quantum mechanical characteristics of the M-Set and it is the intention of this section to begin to underscore the case that

10.1 – the M-Set is the 'original' foundation of Quantum mechanics and, moreover, that

10.2 – the M-Set is the fully renormalised Universal Quantum Field of the M-Set Universe generated spontaneously from the Primordial Quantum Field of the Elemental State of the M-Set by the Trilogy of 'intending' of the ghost force of gravity which manifests in the M-Set Universe as the Dual Gravity Force System together with the complexifications of the Critical Boundary of the M-Set that provides the 'representative basis' for the projective realisation of the Physical Universe.

In the sections ahead all of the properties and principles of Quantum Mechanics will be shown to arise directly and spontaneously from those of the M-Set which, in turn, all relate back to the Defining Property of the M-Set, namely self-referentiality, thence Quantum Mechanics, together with Quantum gravity, will be seen to derive from the symmetry or Holy Grail of self-referentiality.

The M-Set will be revealed, then, as the irreducible, wholly self-consistent foundation of the phenomenology identified as Quantum Mechanics and Quantum Field Theory with the implicitly unifying dual Quantum gravity force system playing the miraculous role of automatically and precisely renormalising the Primordial Quantum Field unto the realm of projective realisation at the Critical Boundary where the apparent or spontaneous breaking of the symmetry of self-referentiality occurs simultaneously with the spontaneous breaking of the symmetry of self-reflectivity, and which we shall unravel further in the sections ahead.

In the M-Set Universe

10.3 – the M-Set is First Cause and even at this early stage of revealing the properties and principles of the M-Set it is already clear that the M-Set is rich beyond imagination with potential which, as we shall progressively discover, can self-project a self-realisation we recognise as 'reality' that is best described at elemental levels by established theoretical systems such as those of the faculty of physics, while, as we shall come to observe in Part II,

10.4 – the M-Set is the origin of all faculties of knowledge because of the property of 10.3.

As this juncture we are focusing on the most primordial level of the M-Set wherein the Quantum Realm resides, bearing in mind here that in projection every level is mixed with every other level to generate the manifold M-Set Universe.

Quantum Mechanics ordinarily finds no explanation by interpolating from the 'physical realm' because the constructs of this discipline and its fuller expression in Quantum Field Theory have no objective basis or foundation there, but rather only generate paradoxes of the most confounding nature that, however, are exactly related to the properties and principles of First Cause of the M-Set.

Stepping off then at the level of the virtual Elemental State of the M-Set, about which we shall continue further discussion in the next section, and with reference also to the nature of this state as revealed up to this stage from the primordial properties of the M-Set, it is clear already that the M-Set naturally and spontaneously generates wave functions or amplitudes which can provide the substrate of the Quantum field and that these arise in this way because of the symmetry of the M-Set, including, of course, the self-reflective symmetry encompassed by self-referentiality, which establishes the primordial reflective fluctuations cohering to the wave functions or the complex amplitudes for all of the possible 'representations' of the M-Set.

In contrast to the language of Quantum Theory, however, we shall exchange the term 'probability' with the notion of possibility because the construct of probability is wholly illusory in the M-Set Approach, as alluded to earlier, while the M-Set is effectively a 'deep well of all possibilities' of itself that ultimately form the basis for self-projective realisation. In truth there are only possibilities, not probabilities, and First Cause spontaneously and freely generates both the possibilities and their 'representations' or wave functions, or amplitudes if you will, through the computational vehicle of the M-Set.

First Cause is not a roulette wheel and, as we shall discover along the way.

10.5 – the M-Set spontaneously generates the mathematical functions and operations that project it.

We shall have occasion to address this property further in Part II of this work while it is helpful to state here that the M-Set can also be regarded, among many things, as the mathematical set because the M-Set spontaneously creates mathematical constructs from its Defining Property which, after all, is not surprising because the M-Set is First Cause, and so it is for the Mathematical formalism of the Quantum Realm too.

We defer now to the previous discussions about 'exponentiation' of the re-presentations of the M-Set of which we shall have more to say again in the section on the M-Set as a Quantum Computer, while noting here that the reflective fluctuations between representations or modes of the freely chaotically convoluting M-Set generate all possible wave functions or representations, or so called amplitudes.

Moreover, in the Virtual Realm of the M-Set there is no dissipation or dissipative interference, as we discussed earlier, while all computations are free and are simultaneously convoluted or folded into each other, which leads directly to the remarkable result that the representations of the reflective fluctuation between all possible modes or Representations of the M-Set are all independent of each other.

Another way to view this is to note that because of the property of non-dissipation of the M-Set the reflective fluctuations between all possible modes or representations of the M-Set are 'preserved' unto projective representation as wave functions or amplitudes so that in the Projective Realm every possible reflective fluctuation is identified by its representation or wave function, or amplitude.

This highly significant property of the M-Set computations, which we shall have occasion to illustrate in other ways in future sections results in the remarkable property of the Projective Realm of the M-Set computations that the representations or wave functions of the re-presentations of the Virtual Realm of the M-Set form a linear space spanned by the independent representations, this linear space being the so named 'Hilbert Space of States' which, physically speaking, is like a linear superposition of wave functions or, if you will, like a pond of non-dissipatively interfering waves. Thence, it is the case that,

10.6 – the M-Set is self-linearising projectively, which, when recalling the wholly non-linear nature of the M-Set, is a remarkable feat indeed while it is directly the case that the foundation principle of the Linear Superposition Principle of Quantum Mechanics is a natural consequence of the M-Set projection.

In effect too, the M-Set computational dynamic spontaneously diagonalises or lines up every conceivable representation of itself into a linear space where each dimension of the space is an independent representation or wave function in projection. From the seemingly unravelable and unimaginably complicated non-linear and chaotic convolutions of the M-Set there emerges an indefinitely dimensioned linear 'Hilbert Space of States' representing as it does now the linear superposition of all possible reflective states for the fully renormalised Quantum Field of the Projective Realm that the M-Set is.

An interesting corollary of the considerations thus far is that the physical manifestation of the Linear Superposition Principle of Quantum Mechanics will be most evident and pure when making observations of the M-Set Universe at microscopic or atomic and subatomic levels which, as we shall learn soon, relate to the Primordial Quantum Field of the Elemental State levels of the M-Set that are relatively uncluttered by the complexifications of Quantum gravity weaving its magic from the primordial skeins unto all systemic levels of projective realisation.

Of course, the Linear Superposition Principle pertains whatever the level of perspective taking in the projective realm, however, in tightly controlled experiments of elementary particle physics at very elemental levels one can approximately focus exclusively on the relevant representational linear space of states of the specified projective state that has only a few dimensions or a handful of representational modes, while evidently the number of dimensions escalate astronomically for all but the simplest systemic arrangements, such as those found in controlled laboratory experiments.

Implicit then to the M-Set self-linearisation is the Quantum Mechanical principle of the Linear Superposition of States which is represented, if you will, by an abstract space called the 'Hilbert Space' consisting of a superposition of independent wave functions or representations of all possible representations of the M-Set.

We observe here too that we are now providing a new interpretation of the wave functions because in the language of Quantum Mechanics these are called 'probability amplitudes' while it is abundantly clear now that these amptitudes have their origin in the M-Set where the notion of probability is anathema. Probability, unlike possibility, is not an actuality of the M-Set or its projections. It is merely a mathematical contrivance relating to the way observations or projective cuts are made.

The currency of the M-Set serving up the observations or projections is possibilities.

The M-Set roots out probability once and for all!

The spontaneous linearisation of the projective realisation of the M-Set embodied in the Linear Superposition Principle can be understood as the M-Set's solution to unravelling the wholly non-linear parallelism of its virtual state in order to be able to effect the self-projective processes described recently and which are summarised in the mnemonic SR4 of self-referential, self-reflective, self-representational self-(projective) realisation.

It is also clear now that the spontaneous projective self-linearisation of the M-Set relies upon the simultaneous interplay of the self-referentially contingent purely chaotic computational dynamic of the M-Set, which will be expounded upon again when we view the M-Set as the Universal Quantum Computer, which it is, and the self-reflectively contingent elements of the computational dynamic, namely the reflective fluctuations between all possible 'modes' or representations of the M-Set.

Thence it can be said here that the spontaneous projective self-linearisation of the M-Set which the Hilbert Space of States of the Linear Superposition Principle represents also represents the simultaneous spontaneous breaking of the symmetries of self-referentiality and self reflectivity. There are, however, a number of subtle stages still to progress through ahead before the story of spontaneous projection at the Critical Boundary of the M-Set can be fully appreciated. Suffice it to anticipate here that it involves the renormalising power of gravity unto the Critical Boundary, the dualism of dispersive and distributive properties at the Triadic causality, and the Quantum mechanical phenomenon of the apparent collapse of wave functions occurring as the M-Set symmetrically or self-interactively projects through measurements of itself.

We have already made mention of the M-Set being viewed as a Universal Unitary Matrix operator, and with regard now to the observations above as well as to the implicitly dual nature of the M-Set, the M-Set in projection can also be understood as the Matrix operator of itself in relation to the 'operand' or Hilbert Space of States with the M-Set thence dualising in projection as both the Universal Wave Function and the 'Universal 'S'-Matrix' in the language of physics.

It is now also abundantly clear from the M-Set perspective that there is no thing without or outside of the M-Set, while the projective realisation is from within the M-Set.

There is no 'without' for the M-Set, and

10.7 – the M-Set projective realisation creates the illusion of reality, with reality being of that which exists, or of that which ex-is in time and space.

There simply is no particle, point, or continuum. These are all singular illusions of projection while the Quantum view of wave amplitudes takes centre stage and the so called wave-particle duality will be seen to arise naturally in the M-Set perspective from the phenomenon of the apparent collapse of wave function when self-referential self-measurements of the wholly non-local M-Set take place at projection. We shall now proceed to explain this very crucial aspect of the M-Set more fully.

Focusing firstly on the wave amplitutes making up the independent axes or degrees of freedom of the Hilbert Space of States, implicit to these amplitudes as alluded to earlier is both the 'redundant information' of the symmetry underpinning the representations that the amplitudes relate to as well as the invariants of the symmetry which will be identified later as Quantum numbers. The purpose here is to highlight again the observation that the wave function involves the functional space of possibilities for the Quantum invariant (compare with the example of the circle being the functional space of the representations or the possible positions of a point circumscribing the same, while the action of the point as it changes positions is associated to a quantity of angular momentum, this also being a 'quantum' or 'amount' that is invariant with respect to the position on the circle where the change takes place because the circle is rotationally symmetric).

We note here for future reference a couple of subtle points implicit to this example, namely that the dualism of structure and function' is allied to symmetry and that the symmetry and dualism lock in a Quantum invariant so that ultimately, as we shall see later, self-referentiality, duality and quantisation are also an important Trilogy of the M-Set.

Now, as we have observed to this point, function is exactly dual to structure at the Triadic Causality of projective realisation of the M-Set, while the M-Set is wholly non-local and, moreover, all projections of the M-Set are also wholly self-reflective in the Projective Realm, the sum result of which is that

10.8 – the M-Set projection is a self-measurement of the M-Set or a Sym-metry of the M-Set. Moreover,

10.9 – the M-Set is the observer and the observed, reflected self-referentially in projection because the M-Set is the 'operator' and 'operand' of itself.

But, every self-measurement of the M-Set is also an interaction of the M-Set with itself, while we know now that the only interactions the M-Set has with itself involve the self-differentiations of the reflective fluctuations between all possibilities of the M-Set Quantum field, thence, every self interaction of the M-Set is a mixture of the reflective fluctuations of the Quantum Field from which it follows that

10.10 – the M-Set self-interactive self-measurements are the projections of the M-Set and these are evidently, from the perspective of the Virtual Realm of the M-Set, occurring now, simultaneously, and in parallel in all possible ways.

Alternatively stated too,

10.11 – the M-Set Universe is the parallelism of all possible self-differentiations of the M-Set.

We observe here in anticipation of our discussion of the M-Set as the Universal Quantum Computer that the mixing of the reflective fluctuations, involving as this does these two simultaneous and spontaneous virtual computational activities of the M-Set, is also the computational basis for the simultaneous spontaneous breaking of the symmetries of self-referentiality and self-reflectively.

As we shall be able to explain in more detail, the mixing of the reflective fluctuations which relates to the 'renormalisation' and the complexification required for the elaboration of a Tower of Mass Effects in projection unto the Critical Boundary of the Projective Realm is simultaneously apparently self-differentiated or divided in every possible way by the reflective fluctuations between the mixed states of the M-Set to generate every possible projective cut of the M-Set, and which projections in turn apparently simultaneously break the symmetries of self-referentiality and self-reflectively. Simultaneously too, the M-Set has self-protectively linearised itself to provide the Quantum basis of all 'observer-other' projections so in this sense the Quantum Realm or the wholly renormalised Quantum fields are like bundle fields, in a mathematical sense, connecting the M-Set Computational Realm with the Projective Realm.

Bearing in mind all of the above, consider here the example of the M-Set making a specific self-measurement and thence a specific projective self-differentiation, while we remind ourselves again that all projections of the M-Set are self-interactive 'observer - observed', 'object - other' self-measurements. Focusing for a moment now on our naïve example earlier of a point on a circle, when a specific self-measurement occurs the space of possibilities apparently collapses spontaneously because a particular possibility is chosen by the projection. Moreover, the apparent collapse is of the space of redundant information or of the space of functional freedom dually associated with its quantum invariant, and as demanded ultimately by self-referentiality in the Trilogy of (self-referentiality, duality, quantisation), as we shall see ahead, while also at the self-interactive self-measurement of projection it appears as if the 'symmetry' has been spontaneously broken.

Stated alternatively, the collapse of wave function phenomenon which is synonymous with the collapse of the space of possibilities is in the Trilogy (collapse of the wave function, spontaneous symmetry breaking, self-interactive, self-measuring, self-projective self-realisation).

The collapse of the wave function phenomenon is synonymous with projection in the M-Set perspective.

In a fuller statement, the collapse of the wave function phenomenon is simultaneous with the apparent or spontaneous breaking of the symmetry associated with the space of redundant information reduced out by the sym-metric computations of the M-Set. The 'redundant information' that is also the 'functional space of possibilities' of the quanta of the associated symmetry generates, in turn, the wave functions or representations which apparently collapse at the self-differential projective cuts of the M-Set.

This is the Trilogy above, in action.

The collapse of wave function phenomenon in the M-Set Approach is also spontaneous in the sense that the wave doesn't retract relativistically. It collapses simultaneously with the self-measurement and self-projection of the M-Set because

10.12 – the M-Set is wholly non-local, while recalling here from the primordial properties of the M-Set that the M-Set has neither space nor time in the physical sense. Indeed, the M-Set creates the illusion of these by the collapse of the wave function phenomenon, as we shall continue to appreciate, and we note here that in effect it is the redundant information of the symmetry of the wholly virtual, non-local M-Set which blows up the illusion of 'space-time' while the collapse of the wave function phenomenon at the self-differentiated interface of 'observer-other' or 'this-that' generates the illusion of objectivity or mass effects, associated as these now are in the structural-functional duality of the Projective Realm with the quanta or invariants of the symmetries which we shall explore further in the next section.

And, because of the wholly non-local property of the M-Set and simultaneity of projection we can understand more clearly why there is only 'now' in the M-Set Universe, bearing in mind the property that

10.13 – the M-Set is all possibilities of itself, now, in all ways and always. By way of clarification of these quite complicated discussions to this point we observe here that in the M-Set Approach the spontaneous, free, self-re-flective fluctuations are the Self-differentiations of the M-Set, and it is by Self-differentiation into the Linear Hilbert Space of States that the M-Set spontaneously unfurls or unravels, or protectively realises itself with the inestimably important caveat, to be elaborated upon in the sections ahead, that it is through the mixing of the Trilogy of 'intending' of all of the levels and orders of the reflective fluctuations of the symmetry of reflectivity that the M-Set spontaneously self-differentiates into the 'this' and the 'other' to apparently spontaneously break the symmetries of self-referentiality and self-reflectivity simultaneously, while at the 'projective cut'of the projective self-differentiation of the M-Set the symmetry of non-locality is also apparently and spontaneously broken in concert with the apparent collapse of the wave function phenomenon. And thence, moreover, we observe also how the 'this' is a reflection of the 'other' across the projective cut, and how for any objective 'this' the 'other' re-flects its sum of histories, now, to perfectly uphold the defining property of Self-Referentiality of the M-Set.

The 'Wave-Particle Duality' phenomenon is now identified with the properties and principles of the projective realisation of the M-Set including, of course, the property of the exact dualisation of structure and function which, at the elemental level of light itself, is revealed as the duality between photons and waves. We shall speak more specifically to this property of Quantum Mechanics in the next section when we focus attention on the Elemental State.

We observe here that 'non-locality', the Wave-Particle Duality and the collapse of the wave function are a Trilogy of aspects of the M-Set which of course are as 'One', self-referentially, and which are inseparably simultaneously superimposed, now, in all ways, always.

Quite generally, Trilogies of the effecting of the M-Set relate to the Triumvirate of the Triadic causality and we can thence say now that

10.14 – the M-Set oneness expresses as Trilogies, even as the M-Set is projecting dually, the Projective Realm is triadically linked back to the M-Set in a trilogy as we explained earlier. And thence of course

10.15 – the M-Set is the origin of Duality while it is very clear now that dualism is incomplete or folded into the greater whole of the Oneness that the M-Set is.

The M-Set provides the context for dualism and so it is, as we shall continue to appreciate, that all the apparent paradoxes of dualism are resolved by the M-Set, in all ways, always, now, while the M-Set is the vehicle for opening our eyes to the realms of Oneness and the Trinity.

Non locality, the Wave-Particle Duality and the 'collapse of the wave function' phenomenon are three apparently paradoxical properties of the Quantum Mechanical Realm because they appear to defy Relativistic dualism and Relativistic Invariance. However, these three properties not only spring naturally from the M-Set but remarkably, as we touched upon earlier, are wholly consistent with the illusion of Relativistic dualism because all projection is borne on light (not just visible light).

We can see more clearly now the origin of the paradoxes as being the triadic collusion between the properties of 'that which is', embodied as they are by the M-Set, and the properties of 'that which is not', embodied as they are by the Illusory Projective Realm of the M-Set, all the while with the M-Set tracking our quest for deeper enquiry and unity.

One can now understand too that it is the collapse of the wave function phenomenon across all orders and levels of the projection which is instrumental in creating the illusion of space, matter and motion or the illusion, if you will, of spontaneous creation through the simultaneous spontaneous symmetry breakings of the M-Set.

It is also at this place of Triadic causality that a deeply profound and subtle effect comes into play. Namely, the force systems make their appearance, quite literally, to exactly uphold the duality of the Projective Realm in the face of the spontaneously vanished or collapsed wave functions and thereby giving rise to field force systems or 'Quantum field force systems', as we shall have occasion to discuss in more detail ahead.

The Virtual and Free Realm of the M-Set projects the illusion of a mechanical world and, in the analogy of the CD-ROM computer game projecting onto the TV screen, the M-Set is all the possibilities of itself now, playing out in the projection of itself.

The quanta associated with the apparent physical force systems will be shown to relate to the invariants of the sym-metries of the reflective fluctuations begetting the representations of the 'collapsing of the wave function', which phenomenon is occurring, in turn, as a result of the projective, reflective, self-interactive self-measurements (or symmetries) of the M-Set about all of which we shall further our discussions soon; suffice it to note here that for the subatomic force systems the distinctions above will be quite clear. However, when looking at complex systemic levels of interaction there is unimaginably complicated mixing of structure and function about which we spoke for the gravity force system, although even this system, like the other elemental force systems, has an elemental level too that we shall investigate ahead.

The Force Field Systems of the Projective Realm of the M-Set are automatically relativistically dualistic and relativistically invariant because all projection is borne on the light-like wave functions or representations of the Critical Boundary of the M-Set, and so it is that the M-Set explicates or unfurls the intimate relationship between its symmetries and the force systems of the physical world.

From a theoretical physics point of view one might identify the M-Set as the Vacuum State of the M-Set Universe embued as it is with the fully renormalised Quantum field that spontaneously creates the M-Set Universe.

There is no beginning and there is no end from the M-Set perspective because these are coincident points on the cosmic wheel of the eternal now, while 'big bang theories' and other dualistically bound constructions are not aware of First Cause and result out of being seduced into following the time lines of the illusions of the Realm of Projective Realisation. It is now more evident from the emerging Quantum Realm of the M-Set that such theoretical excursions will be thwarted by unresolvable singularities of one's own making.

To go deeper into the language of the Quantum Realm and to draw closer to the heart of the formalism of the same from the M-Set perspective we shall need to focus again on the Elemental State of the M-Set and explore further the M-Set computations specific to that level, bearing in mind still that all orders and levels of the M-Set computations occur simultaneously and in parallel, with each level being completely and non-locally 'folded' or mixed into the other unto the spontaneous or apparent symmetry breaking of the projective realisation which corresponds, in turn, to the spontaneous creation of the M-Set universe, wherein every aspect is perfectly reflected in the other, now, in all ways, and always.

11: The Quantum Realm from the Level of the Elemental State of the M-Set

Deferring to the earlier discussions around the emerging properties of the M-Set it became apparent then that there were a number of constraints naturally imposed by the Defining Property of the M-Set on the Elemental State which we wish to explore in more detail here with the observation at the outset that the Elemental State, being confined as it is to the self-referential M-Set and from which the M-Set self-computes, must itself be self-referential because, whatever the order of the computation, it is the case that the manifold of resultant states is in parallel and thence every order must obey the self-referential boundary condition of the M-Set, all the way down to the order of the single Elemental State.

Within the initial condition and the Defining Property of Self-referentiality of the M-Set the Elemental State must be capable of re-presenting itself self-referentially in order to establish the computational process of the M-Set, as outlined earlier when appeal was made to the mathematical notions of purely complex virtual spaces such as Hamilton's three-diementional complex space to provide a guide, while acknowledging here that the latter space is simply exemplary as the lowest dimensional irreducibly self-referential virtual space able to notionally, at least, represent our preliminary proposition for the Elemental State which must be able, moreover, to spontaneously develop representations from its own implicit phasal symmetries.

We recall here that the ability of the Elemental State to re-present itself in a self-reflective way, self-referentially speaking, implies directly that the Elemental State has more than a single mode of itself and that it is not the modes themselves which are being 'represented' by the M-Set but the differences or reflective fluctuations between them, while we make the inestimably important point here that our reference to notional spaces for the M-Set is just 'that' in order that we can illustrate a much more profound truth, namely that ultimately

11.1 – the M-Set self-referentially determines itself regardless of what preconceived ideas we may have for it. Stated alternatively,

11.2 – the M-Set spontaneously represents itself which it freely discovers for itself. Our purpose here continues to be to describe how this self-discovery occurs within the constraints of the descriptive power of the written word.

We have indicated on a number of occasions to this point that the implicit freedom due to the 'redundant information' of the internal symmetries of the M-Set enables the spontaneous development of the reflective fluctuations or the phasal excitations of the Quantum Field of the Virtual Realm of possibilities of the M-Set from, in turn, the self-referentially determined ability of the Elemental State to spontaneously self-differentiate or to develop representations from its own implicit 'phasal' symmetries while, upon projection of the same, the M-Set spontaneously generates matter waves, motion and space by this reflective self-differentiation or division of itself. We recall here too that this is all for free because the ground state(unprojected Quantum Field or vacuum state if you will) of the Virtual Realm of the M-Set is perfectly free with precisely no nett extrinsic variables, where extrinsic variable refers here to a variable taking on a dimensioned value in the Projective Realm such as 'entropy', 'energy', 'charge', 'angular momentum', and 'momentum' while it is the case that the M-Set is teeming with virtual actuality which is revealed by the Divinity or division of unity of projective realisation in a relativistic dualistic way as we have explained it to this point.

Because the M-Set has no nett extrinsic dimensioned variables this automatically ensures the 'universal conservation' of these quantities in the Projective Realm, and thence as a general rule

11.3 – the M-Set only projects universally conserved quantities,

which also follows directly from the observation that the M-Set is wholly symmetrical and can therefore never actually breach this restraint, even though it appears to do so in manifold ways by 'local' perspective taking in projection.

We observe here that the conservation principle strictly applies to the M-Set Universe as a whole and is evidently broken for localised measurements or perspectives such as for variables like temperature and entropy, while 'local' conservation of extrinsic variables in controlled or experimental situations such as for energy and momentum is strictly only approximate, as I shall discuss ahead. Suffice to say here that in such circumstances external or outside intending influences almost identically cancel to all intents and purposes so that 'local' conservation laws appear to apply.

With complicated variables or variables of complexity such as 'distributions' and 'entropy', for example, as well as 'temperature' with its relationship to the 'thermal state' or 'thermal distribution' we shall discover that because

11.4 – the M-Set is completely neutral, all such variables reside in the perfect universal dualism of structure and function of the Projective Realm as clearly the M-Set itself cannot have 'a' temperature or 'an' entropy for example, while we shall continue to appreciate it is necessarily the case too that the Universal Distributive Property of the M-Set Universe is exactly dual to the Universal Dispersive Property across all levels of the manifold M-Set Universe.

We have already met this at the 'neutral' Critical Boundary, while it is now a requirement of the Neutrality of the M-Set that the distributive property of criticality, namely the 'fracticality' ($1/\iota.f$) is exactly dual to the 'dispersive property' 'c' ~ $\lambda\nu$ of the representations or the 'wave functions' of the Projective Realm.

In a naïve way this means that the structure of the M-Set Universe is dual to or inversely related to the function because, from the perspective of 'neutralisation', symbolically $\nu\lambda$ / $\iota.f$ ~ 1 across the manifold of the Projective Realm.

When we state, for example, that the observable manifold universe has a particular temperature what this means in the M-Set perspective is that for 'that' particular projective cut of the M-Set, the manifold M-Set Universe as a whole has a particular neutral dual of structure-function, distribution-dispersion, matter-temperature and so on.

Continuing on now from the above discussions it follows from the parallelism of the M-Set that the Elemental State of the M-Set has no nett extrinsic variable values. Thence, the Trilogy State spontaneously reflectively descending from the Trinity of the Self-conception of the Trinity principle, as was explained earlier, can only contribute to functioning as a possible candidate for the virtual Elemental State if the Trilogy state's representation as the state of virtual spin ½ like phasal degrees of freedom simultaneously reflectively folds or convolutes in parallel with the Trinity which begot it so that a wholly neutral, self-referential entity is formed which in turn, and of itself, is then able to represent projectively or, alternatively stated, is able to reflectively fluctuate through the self-referentially freely allowed computations of the M-Set to set up its own projective realisation that, as we now know, involves the processes of symmetry or self-measurement, including the collapse of wave functions, spontaneous symmetry breaking, and the simultaneous emergence of relativistic dualistic force systems borne on the light-like representations or wave functions of the Projective Realm.

The non-self-referenced or non-self-referential Trilogy state is reminiscent of the virtual three quark state or 'Hadronic' state of particle physics which family of particles includes protons and neutrons. However, in the M-Set Approach because Hadrons (and thence the Trilogy State as we shall see) have nett quantum numbers such as spin and charge in projection the Hadrons must be absolutely conserved. In physical elementary particle terms Hadrons must 'exist' or 'project' as particle-antiparticle pairs that cancel out to light-like wave states or be resident in neutral elements, such as the Virtual Elemental State which can reflectively fluctuate to self-project with light-like representations because, moreover, the neutrality of the M-Set informs us that the M-Set, and thence the M-Set Universe, has no nett Hadron number or number of Hadrons.

The Hadronic-like virtual state of the Trilogy can not qualify as the Elemental State of the M-Set because of its nett Quantum numbers which, as we shall soon learn, also include the invariants of the phasal spins (isospins) of deeper virtual 'internal' phasal symmetries of the M-Set.

Now, as we observed earlier in our discussions of the Trinity Principle of the Self-conception of the M-Set Universe, the purely self-referential and wholly unknowable Trinity State can, because of the inherent freedom of self-referentiality itself, spontaneously self-reflect aspects of itself or spontaneously differentiate aspects of itself self-reflectively provided, of course, this is unto the self-referentiality and the Oneness of the M-Set. We also demonstrated earlier that one possible self-referential reflection is the re-flective self-differentiation of the Trinity state into the Trilogy state reflecting, in turn, via a P(1) or unitary phase degree of freedom with the residual '2-phasal' subset P(2) of the Trinity that begot it, while recalling here that the reflective fluctuations themselves which relate to the 'self-differentiations' or 'changes' of the M-Set must be unitary or 'uni-phasal' changes in the Virtual Realm sense, because the M-Set changes in a perfectly reflective way in order to uphold self-referentiality, and which in the Projective Realm sense equates to the property of Unitarity as we discussed previously.

Evidently too, the Trinity can only spontaneously fluctuate re-flectively in ways that uphold self-referentiality or in ways by which it remains reflectively neutral that will be a very important constraint in guiding us towards a possible self-replicating virtual Elemental State which can form the basis for the M-Set computations.

By way of elucidation of the subtle points above we state alternatively here that the pure self-referential Trinity State of the M-Set which is One can spontaneously self-reflectively self-differentiate self-referentially into the Trilogy State and the reflection' of the Trilogy State in' the Trinity state begetting it whereby, in turn, the Trilogy State and its reflection in the Trinity state begetting it fold-back or reflect upon each other in the Virtual Realm unto the Oneness of self-referentiality itself.

The spontaneous self-differentiation is then seen here to be of those reflecting states that convolute or fold together into the Trinity State.

Miraculously, it will turn out that 'the Hinge' of this reflective folding unto Oneness is none other than the electromagnetic force system of the Symmetry of Reflectively, while the latter symmetry is also the symmetry of the reflective fluctuations unto self-referentiality that begets the light-like self-differentiated projective realisations of the M-Set, and about all of which we shall have more to say soon.

To clarify where the above discussions are leading we shall denote the virtual 3-phase wholly self-referential Trinity State by P(3) which, as we indicated earlier, is also the boundary of our knowledge because in truth this state is wholly unknowable except through the relativistic dualistic self-projection of all possibilities of itself.

The Oneness of the Trinity of the M-Set is the boundary of knowability of the M-Set Universe.

The Oneness of the M-Set embodied in the Trinity projectively realises itself now, in all ways, always.

Focusing now on P(3), we described before how recombinations of the virtual phasal degrees into locked synchronous pairs can create virtual phasal versions of 'spin ½' like objects. For 3 phases the recombinations can occur in 3!/2! ways (in combinatorial notation), or 3 ways, in fact, for the indistinguishable virtual phases of P(3), while in the setting of the self-referentiality of the M-Set the '3 pairs' convolute simultaneously to form a descendant state, namely P(3) ~ P(2) x P(2) x P(2) ~ P(2) x 3 or the Trilogy State that we identify as a virtual parallel state of 3 spin ½ like locked, paired phasal degrees of freedom, or a virtual 'Hadron-like' state of 3 virtual spin ½ entities, which are called 'quarks' in modern physics.

Now, once again, because of self-referentiality any implicit state of the M-Set can convolutionally replicate (fold back into itself or convolute with itself). Another way of stating this is that provided self-referentiality is upheld the M-Set can be in a parallelism of virtual states of itself which are necessarily convoluted or folded into each other to uphold self-referentiality. Alternatively viewed, because of self-referentiality fluctuations of the M-Set between modes of itself must be simultaneously reflective. The reverberations, if you will, of the M-Set are not on a time line but are 'stacked in parallel', now, in all ways, always, so that the M-Set is a non-local, simultaneous parallelism of all possible reflectively fluctuant states of itself.

Therefore, if the Trinity transcends into P(2) x 3 then the virtual simultaneous state involving P(2) x 3 and its reflection in the Trinity begetting it must be self-referentially closed.

The Trinity, if you will, can spontaneously reflectively fluctuate or set up self-referential parallelisms of itself by self-differentiation of itself.

That is, the wholly unknowable self-referential Trinity self-differentiates into a parallel reflective and self-referential fluctuation of itself.

From the self-referential perspective of the M-Set, the M-Set is thence simultaneously both the Trinity and the reflective fluctuations of the spontaneously self-differentiated Trinity.

It now follows that

11.5 – the M-Set and its projective realm, the M-Set Universe, are One in the Trinity, now, in all ways, always.

The first order fluctuation of the Trinity is the virtual elemental state we are seeking and for this state to be self-referential the replicative convolution of P(2) x 3 together with the reflection of itself in the Trinity that begot it must be wholly self-referential and thence neutral too.

The neutrality implies, in turn, the neutralisation of the nett quantum values or irreducible invariants of P(2) x 3 by the convolution of P(2) x 3 with the reflection of this state in the Trinity that begot it. Of course in the virtual self-reflective self-referential world of the M-Set this seemingly complex process is freely allowed or happens automatically and spontaneously, now, while explicating it can appear unusually difficult, but it is not!

By way of further explanation, one possible way the virtual Elemental State can be realised is if P(2) x 3, the 'Hadronic Trilogy', can form a neutral self-referential 'reflective fluctuation' or Quantum fluctuation with its image in the Trinity that begets it, bearing in mind here that the representations of the reflective fluctuations themselves must be able to be represented in projection as light-like waves which are like P(1) waves or unitary phase waves. One such self-referential reflective fluctuant state is the simultaneously reflected self-differentiation of the Trinity involving the coupling of the Trilogy to a pair P(2) of the Trinity through the leftover phase P(1), namely P(2) x 3 ↔ P(1) x P(2) whereby we can say P(2) x 3 re-flectively differentiates 'P(3)' into P(2) x P(1) which simultaneously, in turn, reflectively differentiates the Trinity into P(2) x 3 so that the simultaneous parallel reflective self-differentiation is wholly neutral and self-referential.

I shall explain in more detail how this state is neutral shortly.

Just to recap, we are explicating here how the awesome subtly and flexibility of the symmetry of self-referentiality of the M-Set can enable the completely unknowable Trinity that is One to spontaneously self-differentiate reflectively in order to self-generate fluctuations (Quantum fluctuations) which enable Oneness itself to be projectively self-realised through the Quantum Fields it generates unto all orders of convoluted self-reflections of itself.

Stated more simply and directly now, the wholly unknowable purely self-referential Trinity State P(3) of the Oneness of the M-Set is free to spontaneously self-differentiate by reflectively fluctuating self-referentially between simultaneously mutually differentiated images of itself that convolute unto Oneness, ensuring thereby perfect neutrality, self-reflectivity and self-referentially with self-referentiality itself being the sole property required to ensure all the other properties and principles of this first order self-differentiation of the Oneness of the M-Set.

Because of the Defining Property of self-referentiality of the M-Set the self-differentiation of the Trinity occurs spontaneously while, as emphasised previously, the M-Set automatically explores its own possibilities (deep well of possibilities) to find the grandest from within extension of itself which can in turn be realised because the M-Set establishes a first order basis state, called the Elemental State, to support its self-computed self-projective self-realisations. Our ongoing endeavour is to fully reveal how this 'apparent' miracle occurs.

The properties of the M-Set computations outlined to this point imply that there is only one possible Elemental State forming the basis of the M-Set Universe and we now know this element must be able, thenceforth, to self-reflectively fluctuate to generate the primordial Quantum fields for projective realisation as also described thus far.

Short of having to invoke here the mother of all inventions for the elemental state it will become apparent, and it will be shown to be the case too that the Elemental Computational Unit or State of the M-Set is none other than the purely virtual state of the hydrogen atom, bearing in mind immediately this does not imply from the M-Set perspective that such a thing exists, but rather in projection it appears to exist because of the illusory magic of the Projective Realm of the M-Set which we are about progressively unravelling, layer by layer.

The building block of the M-Set Universe and the doorway to the Subatomic World is the virtual hydrogen atom, the holy water of the Holy Grail that the M-Set is which miraculously, as we shall see ahead, gives up the secrets of the Subatomic World through its ramifications or renormalisations by the gravitational force system that prises open the doorway in the infernos of the Cosmos.

Alternatively stated, in projection it is the gravitational force system that explicates or unfurls the subatomic world of the elemental state, from which it follows that

11.6 – the M-Set unifies the microscopic aspects of the M-Set Universe with the macroscopic aspects unto its own completion.

In the M-Set perspective the Elemental State which will be seen ahead to provide the Fundamental Unit of Computation of the M-Set is also the threshold or the bridge, or the hinge of the doorway between the microscopic and the macroscopic aspects of the Projective Realm of the M-Set we call the M-Set Universe, and the 'Hinge,' in turn, is the Self-differentiation of the Trinity of the Oneness of the M-Set upon which the doorway swings between the macroscopic and microscopic realms of the M-Set Universe, now, in all ways, and always.

We pause here to underline a number of emergent properties of the M-Set before moving forward again, noting firstly now that

11.7 – the M-Set spontaneously self-differentiates and that

11.8 – the M-Set can only self-divide by self-differentiation of representations of a self-referential state of itself.

The wholly unknowable self-referential Trinity of the Oneness that the M-Set is can self-referentially, spontaneously and freely self-reflectively fluctuate in its own image to provide the elemental state or the first order division of Unity (Divinity) of the M-Set computations of self-projective self-realisation, and thence it is the case that

11.9 – the M-Set self-realises through the Divinity of the Trinity of Oneness which henceforth we shall refer to as the Trinity principle. This inestimably profound principle gives rise directly to an equally profound property, namely that

11.10 – the M-Set is self-conceiving or, more generally,

11.11 – the M-Set conceptualises itself as the M-Set Universe which directly leads to the insight of the self-referential nature of all conceptualisation.

The aim now is to continue to explore the natural Quantum Mechanical characteristics implicit to the M-Set while orientating ourselves to be able to investigate in sections ahead both how the M-Set elaborates the Elemental State into the other elements of the periodic table and how the M-Set reveals the subatomic forces.

Returning to the implicit phasal symmetries of the Virtual Realm of the M-Set, we have already discussed aspects of these unto their exponentiation into the self-reflective phasal coherences of the representations or the wave functions of the Quantum Field of the M-Set, while we have also alluded to the M-Set creating the illusion of the physical continuum through the explication of redundant information conveyed by phasal parameters, Θ, in the exponents of the representations or the wave functions of the M-Set projections. The spontaneous process of exponentiation underscores, in turn, the means by which the purely non-linear M-Set unfurls (explicates) itself to form a linear representation of itself. Of necessity to as a result of the nature of exponentiation - exp i Θ, Θ must also convey the M-Set invariants of the phasal symmetries because from the defining property of self-referentiality of the M-Set the 'invariants' are 'dual' to the redundant information defining them and the 'exponent' is the only place they can be explicated to.

A consequence of this is that 'Θ' must now split into a self-referentially imposed dualism consisting of both the invariant and the parametrisation of the redundant information defining the invariant while Θ is to remain dimensionless because the M-Set has no space or time, nor matter or any nett explicit dimensioned value.

Moreover, the exponentiation of the reflective fluctuations, which phenomenon we shall approach from a different direction when we discuss the properties of the M-Set as the Universal Quantum Computer, automatically restrains Θ to be dimensionless.

On the other hand we also have the hindsight of our physical experiences and their symbolic attributions, such as the attribution by the mathematical symbol 'x' to define space together with the construction of a continuous number field spanned by x, and similarly with 't' to define time whilst noting here that these symbols are parametic descriptions of a far more profound and subtle truth, namely that of the redundant information implicit to the symmetries of the M-Set which inflates the Projective Realisation.

A consequence of the above too is that the archetypical invariant which we will persipicaciously denote by 'h' must thence have the same dimensions as the parametric variables, V, of redundant information defining it so Θ must be a ratio like V/h or h/V in order to remain dimensionless.

We also learnt from the nature of the processes of projective realisation outlined thus far and mnemonically encapsulated in 'SR4' that at the heart of SR4 (self-referential, self-reflective, self-representational self-realisation) is the notion of change or exchange of mutual information in relation to the reflective fluctuations of the computations of the M-Set, while the M-Set self-differentiates by way of the reflective fluctuations of itself which, in turn, generates the virtual phasal fluctuations of the Quantum fields of the M-Set.

Change, and thence exchange in the reflective fluctuation sense, occurs between representations or possible parallel modes of the M-Set and we shall now identify the representations with vibrationary modes (or eigenstates in Quantum Mechanical terms) of the M-Set. However, it is not the vibrationary modes individually that play any explicit role in the M-Set projection but rather it is the changes between the vibrationary modes, namely the reflective fluctuations, which are explicated by way of this 'self-differentiation' of the vibrationary modes in all possible ways.

The analogy here is to music where the music is created by the spectrum of differences between notes or tones (vibrationary modes) giving rise to the property that

11.12 – the M-Set projection is the differential of its fundamental re-presentations.

In 'differential geometric' terms of mathematics one can say that

11.13 – the M-Set self-realises in the 'Projective Plane' generated by the differentials of the fundamental representations of itself which incidentally also illustrates the direct association between the mechanisms of projective realisation of the M-Set and the Language of Mathematics such as for calculus and differential geometry of which, of course, the M-Set is the origin because there is no other, and about which we shall have further discussions in Part II of this work.

Just as music is related to the differential spectrum of the fundamental tones or notes, upon borrowing this as an analogy we can make the point here that change equates to differences, while without change creation vanishes. The music stops.

In a Quantum mechanical sense one can view the M-Set as a self-referential potential well of stationary vibrations or stable modes, or eigenstates.

The miracle of the M-Set is that the 'self-referential walls' of the deep well of possibilities also cause the vibrationary modes because of the self-referential self-conception of the M-Set and whereupon, in turn, these vibrationary modes reflectively fluctuate in all possible ways between themselves to quite literally generate the Music of Creation, with the profound consequence too that the frequency spectrum of the creation of the M-Set Universe is manifesting in the cosmic background radiation of every projective cut as discussed earlier.

The Quantum Fields of the reflective fluctuations of the M-Set are thence very complex manifold differentials of the M-Set that support all possible projections of the M-Set unto criticality, while the exponents of the Quantum Field fluctuations are of differences (compare again to differences between notes or tones of music) as a result of which

11.14 – the M-Set projects relatively, not absolutely.

In actuality too, only relative differences have any meaning as there are no absolutes in the M-Set, and just as music derives its meaning from the differentials of tones so it is the case too that

11.15 – the M-Set creates meaning through self-differentiation. Without the M-Set there is no meaning, while it is evident now that meaning is inherently entirely relative and at the M-Set level, reflectively self-referential. And so it is too that

11.16 – the M-Set is the origin of meaning.

There are many important Quantum mechanical consequences emerging here, the first being that that which we call 'quanta' are in fact differences and, moreover, these differences are relatively defined on the background of redundant information or continua. In other words, the continuum is meaningless (redundant information), arising as it does in the context of the M-Set from the redundant information of symmetry. Thence, all meaning is contingent upon differences (quanta) and, relative to the continua, quanta must be finite and discrete. In all generality then the 'self-referential potential well' of the M-Set automatically and spontaneously self-differentiates to generate the 'quanta' of meaning in the M-Set Universe, from which it now follows that

11.17 – the M-Set spontaneously creates a quantised basis for the projective realisation, because continuum is the death of meaning. And so it is too that

11.18 – the M-Set is the origin of quantisation.

An extremely profound consequence of the M-Set quantisation and of the M-Set generally is that in projection one never sees individual modes or states, or particles or atoms, or any thing because these never actually exist. In truth we only see the exponentiated reflective fluctuations between the states or representations begotten by the symmetries of the M-Set because that is all there is. Just a 'light show' that projects an image of things or builds up a picture of things.

Meaning and quantisation are one and the same thing in the M-Set approach.

Projection, in truth, is an enlightment or a projection borne on light (not only visible light) generated by the exponentiation of the reflective fluctuations of the M-Set over all orders and levels of its representations or possible states, while the 'deep well of possibilities' is unimaginably vast and includes not only every possible 'intention' of gravity but the Subatomic World as well.

In all generality then the M-Set self-referentially reflectivity represents itself and thereby self-projectively self-realises or represents itself in all possible ways or, to state this in more mathematical language,

11.19 – the M-Set spontaneously generates every possible representation of its own symmetries.

The M-Set is thence self-representing, while mathematically speaking, as we shall discuss in broader detail later, the M-Set is the mathematical construct for representing all symmetries subsumed by self-referentiality which, because the M-Set is 'that', implies too that

11.20 – the M-Set is the universal mathematical Set.

Because of self-referentiality and the resultant properties of projection the possible M-Set representations are like an orchestra with the orchestra being viewed here as the potential for creating music while for the M-Set the defining property of self-referentiality directs that the M-Set can only create harmony because this is assured by the criticality of the conduct of projection at the Critical Boundary of the M-Set where, consequent to the nature of criticality physically speaking, the distributive property of the dummy variables of the redundant information we call time and space is fractal (~ 1/ι.f), as indeed is the frequency distribution of meaningful music or harmony, and which distribution in turn is simultaneously exactly neutralised by the dispersive property 'c' of the 'light-like' representations (namely $c \sim \nu\lambda$) on which the projective realisation is borne at criticality.

We emphasise by way of reiteration here that the neutrality property of the M-Set consequent upon the defining property of self-referentiality of the M-Set together result in the Projective Realisation being wholly critical or on the edge of chaos, not only because the M-Set is pure chaos and is the boundary of itself but also because, simultaneously, the fractal distributive property of the dummy variables of redundant information at criticality (which we shall identify ahead with variables of space and time or the continua of the Projective Realm) is exactly neutralised by the dispersive property of light at the Critical Boundary of the M-Set that the M-Set is.

Moreover, because the M-Set is the Critical Boundary, the distributive (structural) and the dispersive (functional) properties at this boundary must be exactly dual so that in effect these properties of the Projective Realm dually neutralise across the manifold M-Set Universe as a whole.

Thence, what we see in projection is a harmony of unimaginable profundity based upon the reflective fluctuations between the representations of the M-Set somewhat like the tonal differences of an orchestra of instruments creating a symphony with every possible performance being a projective realisation.

Now that we recognise the M-Set projects in a quantised way we shall continue on our journey to uncover the insights the M-Set can reveal about the nature of quanta and subsequently Planck's Constant and the Heisenberg Uncertainty Principle as well as the operator calculus of Quantum mechanics. However, there are several general foundational insights we shall need to draw into the discussions as we proceed here in order to confirm the emerging truth that the M-Set is the origin of Quantum Mechanics.

We have already observed the computational roles of representations and reflective fluctuations in projection and one of our tasks soon will be to further clarify the nature of the exponents of the critically exponentiated reflective fluctuations of the Quantum field of the M-Set as we continue this journey while we note here that up to the critical SR4 projection the 'substrate' is the 'Virtual Realm' or 'deep well of possibilities' of the M-Set and thence too, depending upon the point of view or perspective of the observer will depend the level(s) of the reflective fluctuations that are substantively represented in the realisation of the projective self-differentiation of the M-Set.

Also, because the projection is quantised by virtue of the recent observations, and because the M-Set has no scale or dimensioned value, the archetypical quantum 'h' must divide the continuum, thence Θ - V/h while, moreover, because the M-Set has no nett dimension 'V' and 'h' must have the same dimensions to ensure that Θ is dimensionless which is also automatically assured by the exponentiation of the self-linearisation of projective realisation as we explained recently.

More generally now, because the M-Set is self-referentially closed it follows that

11.21 – the M-Set projections are complete solutions of the M-Set which also means that there are no cut-offs or left overs, or remainders and that the substrate or solution of the M-Set is perfectly reflectively folded (completed) upon itself in relation to every projective cut or projective self-differentiation of the M-Set. This implies also that

11.22 – the M-Set projections are fully saturated from which it follows that

11.23 – the M-Set solutions are perfect meaning in turn that every thing is wholly accounted for 'per' its 'effect' 'reflectively', from possible solution to possible solution, projectively speaking.

It is also the case that the actual change which occurs within the virtual realm of the M-Set and which is represented by matrical operations in the

Hilbert Space of States is motivated by the Trilogy of (intending, extending, subtending) that maximally reduces out redundant information by way of the reflective fluctuations of the M-Set to maximise dual structure and function in the projective solutions.

But, the M-Set itself has no nett change and thence, as is implicit to the symmetry of reflectivity of the M-Set, to every change there is an equal and opposite (reflected) change. However, physical change only becomes apparent at projection when a perspective or a point of view is taken that reveals the duality between observer and observed which, in turn, are mutually reflected exactly in the whole of the M-Set. Thence too for every action in projection there is an equal and opposite reaction, or alternatively stated, every action is reflected or reacted in the M-Set Universe. However, because of the enormous mixing of the sub-tendings into 'intendings' and 'ex-tendings' at complexified systemic levels to form complicated dual structure and function the actions and reactions are apparently 'spread out' or dispersed, or complexified themselves.

For example, if a complex organism such as a human being makes an utterance the M-Set Universe will answer (react) in unimaginably complex ways, and, by the properties of the M-Set there will be a per-effect reflection, or perfect reflection in the Projective Realm sense because the M-Set Universe is self-referentially closed. This underlines the perfectly saturated nature of the projective solutions and underpins the Law of the Projective Realm of the M-Set that for every action there is an equal and opposite re-action simultaneously unto all levels of complexity of the M-Set Universe.

We reiterate here that the only aspects of the M-Set which are projected are the changes between the possible representations of the M-Set itself, thence the reflective fluctuations forming the Quantum field platform for projection will be the focus now for comprehending more clearly the nature of the M-Set action.

Taking the Elemental State model of the virtual hydrogen atom in-volving, as we explained earlier, the P(1) uniphasal linkage between the Trilogy State P(2) x 3 or the virtual Hadronic state and the P(2) virtual spin ½ state of the remaining 2 phasal degrees of freedom of the Trinity state, P(3), that self-referentially reflectively self-differentiates in its own image in the way we have recently explained, we now need to explore what the differences between one representation and any other are for the

Elemental State while noting here that the representations or modes themselves actually remain forever virtual.

The first step is to recall again that due to the perfect, free, self-referentially reflective 'phasal' symmetries of the M-Set, the M-Set can develop stationary states or virtual vibrationary modes of itself which, in terms of the classical physical Quantum Mechanical Hydrogen Model, equate in projection with the so called 'eigenstates' or the imagined (imaginary) stationary, stable, closed orbits of an electron around a proton. But, we now know that in truth there is no such thing as a proton or an election, or any thing for that matter. There are only projections of the reflective fluctuations between virtual states or possibilities of the Virtual Realm of the M-Set that give the appearance that 'things' exist.

The eigenstates never exist. The exponentiated, projected reflective fluctuations between representations or eigenstates are identified here as the so called complex probability amptitudes in the language of Quantum mechanics or as the independent representational axes of the Hilbert Space of States which we spoke of earlier.

It is important to pause here and acknowledge that the concept of a complex probability amptitude which is so alien to a concrete physical view of reality naturally arises from the spontaneous self-referentially generated reflective fluctuations of virtual phasal representations of the M-Set. To state this alternatively, the possibility of probability amptitudes is a consequence of the Defining Property of the M-Set, namely self-referentiality, and all representations of the M-Set are like 'eigenstates' of the M-Set that are being identified here as unique, closed, virtual vibrationary modes of the M-Set.

Associated with the reflective fluctuations or closed loop transitions from representation A to representation B and back again is a 'mutual exchange of information' which is also maximal for the AB-BA reflective fluctuation, and similarly for the BA-AB fluctuation because the mutual information can not be greater for this circumstance while it is also true it cannot be less. One might characterise this by stating that the mutual information exchange is the 'maximal' information overlap for the 'AB' re-flective fluctuation and is the 'minimal' information required for the 'AB' reflective fluctuation to occur while astonishingly, as we shall soon come to terms with, the above also alludes to the truth that the fundamental 'change' or action associated with the Mutual Information of the reflective fluctuations of the 'representations' of the Projective Realm of the M-Set is a unique discrete constant.

Also, in order for the mutual information to be meaningful it must be discrete or quantised as we recently discovered, while we also know now that in the absence of redundant information or, in the projective sense, in the absence of the continuum discreteness is undefined, while the converse is true as well

Moreover, we also now know that the only quantities which can be defined by a continuum are the 'invariants' dually begotten by the symmetries of the invariants, and because the M-Set dualises in projection into the redundant information and invariants' of its implicit symmetries in all generality it follows that

11.24 – the M-Set only projects quantities or structures that are related to invariants of its internal symmetries.

Given that the M-Set is all that is implies in turn that in the Projective Realm all quantities or structures are related to discrete invariants of symmetries begetting dual functionally related spaces of the redundant information of the symmetries, so 'apparently', then

11.25 – the M-Set projects 'quantised' structures and continuum functional spaces arising from the internal symmetries dually begetting them both.

From the above discussions it also directly follows that whenever there is a continuous variable as a result of the projection of redundant information of the M-Set there is also a discrete quantity or invariant associated with the symmetry from whence the dummy variable of the redundant information arose while, furthermore, because the M-Set has no nett discrete values these discrete quantities are universally invariant and conserved (no nett universal value).

Thence it is the case now that

11.26 – the M-Set is the origin of the basis for the relationship between symmetry and conservation laws in projection. while it is highly significant to note here that although the conservation laws are universal they might apparently apply locally too if external influences cancel out or are regarded as negligible. For example, when measuring the momentum conservation of billiard balls on a table we are ignoring all other gravitational influences of the Trilogy of intending of the rest of the universe on the measurements, both macroscopically and microscopically relative to the level of observation of balls on the table, but in truth all the levels are there but they effectively reflectively cancel except for the level of observation of the projective cut, which in this case is billiard balls on a table. This is a deeply subtle and important aspect of observation and measurement in the M-Set approach because we are seduced by this into viewing 'reality' locally when in actuality there is only Oneness.

From the discussions thus far we have also come to know now that extrinsic or projected quantities dually related to continua or redundant information must be quantised and thence the M-Set informs us that quantities such as energy, momentum, angular momentum, and spin must be quantised because they are all dually coupled to continuum dummy or parametric variables which express their dual function in projection. This in turn draws us back to the language of Quantum Mechanics in which framework the quantity we call energy is dually coupled to the dummy variable we call 'time', and the quantity of momentum to the dummy variable 'space'.

In truth, the notion of Energy, for example, is only dually defined by Time, and Time is dually defined by Energy. They are never separate notions, but rather they are relatively dually defined notions as we also alluded to for properties such as distribution and dispersion to this point.

In the M-Set Approach then, there is no such thing as time or energy alone. Time or energy do not exist as separate entities and time and energy are mutually dually defined.

The dualities of time and energy, and of space and momentum, are among the most fundamental of the dualities of the Quantum Realm of the M-Set, while we need to remind ourselves that the complete interdependence of dualism means that how we choose to represent space defines how momentum is represented, for example, and vice versa.

There are no absolutes in the M-Set Universe. The definitions of quantities, qualities and values are entirely relative. There is no space alone. Space only appears to exist because of motion (momentum) and vice versa; momentum only appears to exist because of space. Neither exists alone while their apparent existences are a trick of the contingency of dualism by the mechanism of the M-Set projection of the Triumvirate of Triadic Causality at the Critical Boundary of the M-Set that the M-Set is, and as was explained earlier.

We know also from the Defining Property of Self-referentiality of the M-Set that this property implies operator–operand duality, or in other words that

11.27 – the M-Set is both the object of operation of itself and the operator of operation of itself.

The operator–operand duality is the most profound or primordial duality of the Quantum Realm of the M-Set with enormous ramifications, the first being that it directly follows all the M-Set dualities of the Quantum Realm must be of this nature too because

11.28 – the M-Set is the origin of duality which is also the case because the M-Set is First Cause through the Trinity Principle and thence the dualities of x and p (space and momentum) and t and E (time and energy) are, fundamentally, operator–operand dualities too as we shall show below.

In the Projective Realm the above properties of the M-Set provide the basis for the Quantum Operator Calculus that leads directly to the Wave Function Equations of Quantum Wave Mechanics and which, in turn, sets up the whole domain of analytical differential calculus and thence mathematical analysis in the continua of the dummy variables of the redundant information of the symmetries of the M-Set to underline the truth, once again, that the M-Set is the progenitor set of mathematics because mathematics does not 'exist' *a priori.*

The language of mathematics is implicated in the projective expression of the M-Set and analytical 'calculus' is founded on the Primordial Duality of the M-Set.

Deferring to standard and historically accepted symbolic representations of the derivative language of Mathematics, for the projective realm of the M-Set we can now make the 'familiar dual' assignments of $p \sim \delta/\delta x$ and $x \sim \delta/\delta p$, and similarly $E \sim \delta/\delta t$ and $t \sim \delta/\delta E$ as a consequence of the fundamental operator-operand duality of the M-Set.

Interestingly, as an example of the 'multiplicity' of ways in which the M-Set is self-consistently projectively expressing unto the Defining Property of Self-Referentiality we shall soon observe ahead that the primordial operator–operand duality of the M-Set automatically requires the representations of the projecting M-Set to behave 'functionally' or 'mathematically' like exponential wave functions in order for the dual assignments to be realised in the projective realm.

Put more simply, the fundamental operator-operand duality of the M-Set also demands exponentiation of the reflective fluctuations in projection.

Self-consistently too from 'the 'Projective Realm' and analytical mathematical perspectives, the operator-operand dualities of the

Quantum Realm Operator Calculus generate wave function solutions or representations from the differential equations that arise directly from the dual operational nature of extrinsic variables, such as is the case for Schrodinger's equation, for example, which emerges from the classical mechanical definition of energy, $E = p^2/2m + V$ (potential energy), when the extrinsic variables are replaced by the dual differential operators.

We shall now look to the M-Set to see what more it has to tell us about the nature and quantisation of Mutual Information that is so central to the projective dynamic of the M-Set which will lead directly, in turn, to the notion of action in mechanics as well as the Heisenberg Uncertainty Principle in concert with the fundamental operator-operand duality.

The implicit and fundamental operator–operand duality of the M-Set as a whole is extremely profound and subtle and implies directly as a consequence of the M-Set's virtual, free, self-referential state that the M-Set can spontaneously change in every possible self-consistent symmetrical way which is also expressing as the spontaneous 'dynamism' or 'animation', or indeed vitality of the M-Set Universe into which the Fundamental Duality emerges as the Quantum Mechanical Dualisms of The Projective Realm. Therefrom too emerge the elaborations of the structural–functional duality of the Projective Realm by the renormalising 'ghost force' system representing as complexifications of the Primordial Quantum Field about which we shall have occasion to further discuss in the section on the M-Set as the Universal Quantum Computer.

We make the observation here, however, that the M-Set is able to change in all possible ways while maintaining its virtuosity and pristine properties because it can spontaneously, reflectively, fluctuate, whereupon in turn, the computational power of self-referentiality vastly compounds the representations or possibilities of the Virtual Realm through replicative convolutions, as previously discussed at some length, while we now understand in a more profound way from the recent discussions above how the notion of 'change' is linked to duality and thence to projective realisation.

It is also evident from the very nature of self-referentiality itself and the recent discussions that any change in the M-Set can only occur dually, or to state this another way,

11.29 – the M-Set changes are dually symmetric. Any other change would not be consistent with self-referentiality, and thence the Mutual Information of the reflective fluctuations of the M-Set is both quantised or discrete and dual.

From the fundamental physical dualities of the Quantum Mechanical Realm above it is evident now that the extrinsic variable 'V' of the dual Mutual Information implicit to $\Theta \sim V/h$ must have the dimensions of the 'duals' xp or Et, or combinations thereof, such as xp-px, all of which carry the physical dimension of angular momentum, and therefore because Θ is dimensionless the quantity 'h' which we now identify as the discrete quantum of Mutual Information must have the dimension of angular momentum too.

Focusing again on the fundamental operator-operand duality.

11.30 – The M-Set only transforms or changes into itself. It therefore follows that

11.31 – the M-Set changes are dually reflective and these changes only become evident through the Divinity or 'division of unity' of the M-Set of the SR4 projective realisation when the Primordial Duality of the M-Set is explicated or unfurled.

At the non-renormalised Quantum Field level we are in the realm of the 'Classical Quantum Mechanics' which one might identify as the 'dual relativistic M-Set projection of the fundamental reflective fluctuations or 'changes' of the Primordial Quantum Field of the Elemental State, which in current physics parlance would amount to the Quantum Mechanical properties of 'the non-renormalised wholly virtual hydrogen atom', the Doorway to the Subatomic World and the irreducible Computational Unit of the Cosmos of the M-Set Universe.

Now, it also follows from our discussions to this point of the fundamental operator–operand duality that under the umbrella of the defining property of self-referentiality

11.32 – The M-Set is dually symmetric

or implicitly harbours the symmetry of duality, while it also follows directly here from both the fundamental duality and from the perspective of the wave functions or the representations of the Projective Realm of the M-Set that, mathematically speaking, the fundamental self-transformations of the M-Set are phasal changes, as we initially surmised and which leads directly to the profound property that

11.33 – the M-Set changes are phase changes when referenced to physical reality, and which is also in keeping with the property that

11.34 – the M-Set projects 'critically' or at the edge of chaos.

Physically speaking this means that the M-Set is constantly at a critical point of phase changes as a whole while the properties that

11.35 – the M-Set has no nett dimension and

11.36 – the M-Set has no absolute scale or absolute measure of itself imply that

11.37 – the M-Set is scaling invariant, and all of which properties are now self-consistently seen to relate to the critical point of projective realisation of the M-Set as a whole.

It is crucially important to bear in mind here that local perspectives of the projective cuts will result in the spontaneous or apparent breaking of criticality and therewith the spontaneous or apparent breaking of the symmetry of scaling invariance, while we are now beginning to realise more and more how all of the illusions of reality are created by the spontaneous breaking of the symmetries of the M-Set subsumed by the overriding symmetry of self-referentiality.

We are continuing to proceed to unravel how in truth it is the local perspective taking of the self-measurements or sym-metry of the M-Set which spontaneously creates the 'matter realisation' of the Universe wherewith all the symmetries of the M-Set simultaneously spontaneously and apparently break, while we now know that it is the Oneness of the M-Set which makes this possible.

I note here that our initial intuition about the phasal nature of M-Set self-transformations now automatically emerges through a simultaneous parallelism of self-consistent properties of the M-Set unto the Defining Property of Self-referentiality, and all of which lead us to uncover the remarkable property that

11.38 – the M-Set projections are 'critical' phase changes of the M-Set, now, in all ways, always.

Therefore, the manifestations which we identify as being of Quantum mechanics in the Projective Realm of the M-Set Universe are also dualistically critical.

Miraculously, it is the changes themselves that apparently or spontaneously break the symmetries of the M-Set, and about which we shall have much more to say in the sections ahead, while reminding ourselves again that at the Triadic Causality the Projective Realm takes on the relativistic dualistic properties of the representations of the M-Set which, as we shall see shortly, also confirms that

11.39 – the M-Set is the origin of Relativistic Quantum Field Theory.

Change is also self-differentiation of the M-Set as we already know, hence

11.40 – the M-Set self-differentiation is dualistic.

Making the associative link now between the phase changes of the Virtual Realm of the M-Set, relating as they do to implicit phasal symmetries such as those implied by the various P(Θ) of the Virtual Hydrogen Model above, and the Projective Realm operations of rotation which are angular momentum operators such as 'xp-px' for a single phase symmetry P(1), and noting further from the fundamental dualities of Quantum Mechanics that the product of fundamental dual variables such as x and p have the dimensions of angular momentum or action in the classical Hamiltonian dynamical sense we then begin to see the emergence of a connection between the reflective fluctuations of the M-Set Quantum Field, relating as they do to reflective changes between representations of the M-Set, and the Operator Calculus of Quantum Mechanics in the Projective Realm, acting as it does on the phasal waves or representations of the Quantum Field mandated by the M-Set with the differential operators of Quantum Calculus bringing down physical (albeit quantised) quantities into the projective space of redundant information of the symmetries of the M-Set which is reduced out, in turn, by the M-Set computations based as they are on the self-differentiations of the reflective fluctuations or dual self-transformations of the M-Set.

Evidently then the Triadic Causality of the M-Set involves a parallelism of simultaneous symmetrically related processes whereby the reflective self-diferentiation of the M-Set is orchestrated upon its own implicit, virtual vibrationary modes and parallely projects into the Projective Realm in a fashion described mathematically by 'differential geometry' and 'continuum analytical calculus' begotten by the M-Set. More self-evidently now, the M-Set is seen to be acting as the progenitor of mathematical constructs while we know now that none exist without the M-Set.

We shall continue on our way again to explore how the M-Set is able to naturally projectively complete the entire cast of Relativistic Quantum Mechanics from its Defining Property, while we would like to draw attention here to the property that

11.41 – the M-Set reflectively self-differentiates,

as we explained it earlier by the Trinity Principle of the Self-conception of the M-Set which leads to the inestimably profound property for all phenomenology of the Projective Realm, namely that

11.42 – the M-Set Universe is reflectively dualistically self-discriminating, in all ways, always, now. So, in effect, all physical measurements, observations, perspectives, points of view, ideas, theories are self-discriminations (or in the Virtual Realm of the M-Set, self-differentiations) which result in the spontaneous creation of the Observer-Other or the This-That reflective dualisms of the Projective Realm, as we shall discuss again in Part II of this work, the point we wish to emphasise here being the deeply subtle relationship in the M-Set approach between measurement (observation, theorising...) and spontaneous creation! Therefore, the M-Set completes the apparently paradoxical nature of measurement or observation of the Quantum Realm.

The reflective phasal fluctuations characterising the Quantum field and represented in the Projective Realm generically as the phasal wave function $\exp(i\Theta)$ form the link between the virtual computations of the M-Set, which can be likened now to a spontaneously active computational support, and the non-active or descriptive analytical mathematics of (in this example) differential geometry in the Projective Realm.

It is a highly significant point to reiterate here that the M-Set is First Cause and that Mathematical descriptors, language and constructs are actually projectively spontaneously created by, or are a consequence of, the M-Set virtual computational dynamic while, in truth, no mathematical object actually exists because the dualistic projection is an illusion of the M-Set and thus mathematical objectivity is an illusion too.

Mathematics is a descriptive scaffolding constructed by the M-Set around the dualistic projections of the M-Set because the M-Set is that which gives meaning and life as well as First Cause to these descriptions.

In truth, moreover, as we shall come to see in Part II of this work, the function of thinking of the brain relates in the Virtual Computational Realm of the M-Set to the reflective fluctuations of the Trilogy of intending at the highest levels of complexification of the manifold of the fully renormalised Quantum Field of the M-Set which, remarkably, also generates our mathematical descriptions so that in effect, all phenomenology of the Projective Realm of the M-Set is subsumed by the Trilogy of intending of the Defining Principle of Self-referentiality of the M-Set.

It is also clear, then, that the descriptors or constructs of the Projective Realm are not unique and that other variables or mathematical constructs and tools might sometimes be more parsimoniously efficacious so that when we defer to mathematical symbolism we are in fact also deferring to conventions of historical antecedence rather than to some irreducible truth, while by our recent discussions it is evident that conventional constructive mathematical characterisations of Quantum Mechanics are both wholly supported and generated by the M-Set.

The passages above also explain why the written word rather than a mathematical treatise is the road we have chosen to reveal the M-Set.

Alternatively stated, the M-Set is not a mathematical object. It is no thing. It is First Cause, including of all mathematical descriptions of the language of mathematics. And so it shall become known too that Mathematics is but one example of the languages of the M-Set Universe, while the written word can communicate the simultaneous parallelism of processes behind meaning more broadly, even unto assisting us now in elucidating the meaning which the M-Set Approach is bringing to bear on the foundational properties of Quantum Mechanics and Quantum Field Theory.

And yet there are those who will say that

11.43 – the M-Set is the mathematical set which can be said too because

11.44 – the M-Set completes every perspective and point of view.

Focusing now on the phase 'Θ' of the representations or wave functions of the projecting M-Set, which is symbolically related to the reflective changes of the M-Set begetting the same, we remind ourselves at this point that the Defining Property of Self-referentiality of the M-Set, together with the projective dynamic consequent upon that, have both determined that 'Θ' is dimensionless, implicitly dual and quantised. Moreover, the duality must reflect the fundamental operand–operator duality of the M-Set as well as the duality between structure (quantisation or discreteness) and function ('redundant information' or the continuum).

We also know from the M-Set properties of projection that for the mani-fold M-Set Universe as a whole the dispersive property of the M-Set phasal fluctuations must exactly neutralise the critical 'distributive' property of the 'space' of projection which 'neutralisation' property of the Projective Realm of the M-Set is symbolically represented as $\lambda.\nu/\iota.f \sim 1$ as we discussed previously.

In a projective realm spanned by space and time the critical distribution of 1/length x 1/frequency is exactly neutralised by the critical dispersion $c \sim \lambda\nu$ which we shall come to see is what the velocity of light actually is, bearing in mind again that at the Critical Boundary (and 'c'-boundary or light causality boundary) of the Projective Realm of the M-Set there is a perfect (per-effect) collusion between the 'apparent' projection and the isness of First Cause of the M-Set which, while remaining faithfully beholden to the symmetries of the M-Set, creates the appearances of spontaneous symmetry breaking that are in turn the Dualistic Projections of the M-Set about which we are preparing ourselves for a more definitive picture soon.

It is timely now to point out that the terms neutralise and 'balance' when used with reference to the dispersive-distributive properties of the M-Set Universe refer to the upholding of the scaling symmetry property of the M-Set in the Projective Realm because if the effects of these properties of the M-Set Universe as a whole did not exactly cancel in projection then there would be remnant scales contradicting the scalelessness of the M-Set. Consequently there are some conclusions we can draw directly from the scaling symmetry property of the M-Set, the first being that in order for the scaling invariance of the M-Set to be upheld as a whole all continua variables must be divided by scales of themselves in the sense of 'x/λ' with the continuum variable 'x' being divided by the discrete or scaled length 'λ' so that the ratio x/λ has no absolute scale. Only relative scale is implicit here and for self-consistency all continua variables in Θ must be implicitly scaled by discrete and similarly dimensioned quantities.

We reiterate here that it is a consequence of the scaling property of the M-Set that the distributive and dispersive properties thereof must exactly neutralise each other in projection otherwise extant scaling properties would occur for the Projective Realm we call the M-Set Universe, while clearly the scaling symmetry appears to be broken when making observations or perspective taking locally which leads in turn to the phenomena associated with localised projective self-measurements or perceptions of the M-Set, such as for all of the perceptions associated with apparent thermodynamical phenomena because in truth

11.45 – the M-Set Universe is a perfect critical dynamical equilibrium now, in all ways, always, with exactly no nett entropy, thus reaffirming once again that for whichever projective cut of the M-Set is taken as a whole the dispersive property of the M-Set Universe must exactly balance the distributive property as we discussed earlier.

We need to emphasise these points above for our ongoing discussions while we recall here that it is the gravity force system of self-referentiality itself which is responsible through the Trilogy of 'intending' of the reflective fluctuations of the M-Set for the dispersive-distributive properties of the M-Set Universe so that thermodynamical phenomena are local perspectives of the gravity force system, projectively speaking.

In other words, thermodynamics generally is an illusion of perspective taking in the Projective Realm of the M-Set underscored by the gravity force system of self-referentiality itself.

The laws and conclusions of homeostatic thermodynamics are not true for the M-Set Universe as a whole. They are illusions too that like all of the illusions of the Projective Realm relate back ultimately to the simultaneous spontaneous breaking of the symmetries of self-reflectivity and self-referentiality about which we shall have more to say soon.

We remind the reader here that the actions of making observations or perspective taking are synonomous in the M-Set approach with self-differentiatial self-projections of the M-Set which, in turn, spontaneously project the perfectly (per-effect-ively) reflective dualities of the 'observer-other' or the This-That across all levels of the manifold M-Set Universe, and this also underlines the deep subtlety of the 'self-measuring' self-realisations of the M-Set underpinned by the quantised maximal mutual information exchange information processing of the reflective fluctuations of the Trilogy of 'intending' unto the critical boundary of the Projective Realm.

Now, because the M-Set is non-local it also follows directly that in projection the critical dispersive property 'c' which must be implicit to 'Θ' must also be a constant relative to the space of projection, or, stated alternatively, the universal critical dispersive property 'c' must be a constant of the M-Set Universe because of the symmetry of non-locality of the M-Set.

In physical projective terms the velocity of light is a constant of the M-Set Universe because of the non-locality of the M-Set and this so called velocity that is, in fact, a constant of dispersion is also the critical dispersion which exactly neutralises the fractal distributive properties of the space of the manifold Projective Realm of the M-Set computations to maintain the exact neutrality of the self-referentiality of the M-Set in the Projective Realm as a whole.

In a deeply profound way to be elaborated upon further in the sections ahead the 'light-like' representations with the dispersive property 'c' simultaneously conspire with the critical distributive properties of the chaotic computational dynamic of the M-Set begetting them to maintain the perfect neutrality of the self-referentiality of the M-Set which, of course, also automatically upholds the scaling symmetry property of the M-Set in the Projective Realm.

We also now know that associated to the dummy variables 'x' and 't' are discrete quantities because by the duality requirement of the M-Set 'x', for example, has no definition without a dual discrete quantity, which we shall denote by 'λ', that is a discrete length or wave-length as it will turn out to be, and similarly for 't' there is a discrete time interval 'τ' such that a discrete frequency 'ν' can be defined as $\nu \sim 1/\tau$. But from our discussions to this point we know that because of the scaling symmetry property of the M-Set 'λ' and 'τ' must also be scaling divisors of the continuum dummy variables x and t for space and time that are implicit in Θ respectively in order to uphold the no absolute scale property of the M-Set. We know too that the dispersive constant 'c' is the product of a discrete wavelength λ and a discrete time interval $c \sim \lambda/\tau \sim \lambda\nu$. It follows then that because 'c' is the universal dispersive constant of the Projective Realm the scaling divisors 'λ' and 'τ' can only be the wavelength λ and time interval τ of this universal constant of dispersion, c.

We can state then that the scaling divisors λ and $\tau \sim 1/\nu$ of the 'space-time' M-Set Universe 'multiply' to the Universal Critical Dispersive Constant $c = \lambda\nu$ for all λ and ν, and not just for 'visible' light.

The importance of these observations in understanding universal gravitational and thermodynamical phenomena is enormous, while it is also the case now that 'c' is the Universal Scaling Constant of the M-Set Universe at the Critical Boundary of the M-Set that the M-Set is.

The M-Set approach is providing us with the inestimably profound insight here that space and wavelength, for example, are dualistically related in projection. That is, space of itself is illusory. There is no such thing as space. Space and similarly time is effected in a discrete way in projection, although in macroscopic or classical mechanical settings this is hardly noticeable until one descends to microscopic or Quantum mechanical levels, while the so called velocity of light, c, has the profoundly important roles as both the Universal Scaling Constant and the Universal Constant of Dispersion in the M-Set Universe.

Space and time are dual to the discrete effects defining them. You can not separate space from spatial effects. Continua is an illusion of projection and the dispersive property $c \sim \lambda \nu$ of space-time, which behaves like a velocity in the Projective Realm, is a constant of the M-Set Universe as a whole. Moreover, because of the dual relationship of 'c' to the space of projection classical relativistic invariance is automatically a property of the Projective Realm of the M-Set. Relativistic invariance, then, is a consequence of the universal constancy of 'c', which is both the constant of the dispersive property of the illusory space-time continuum and the implicit scaling properties of the M-Set. It now follows therefore that 'Θ' must be a relativistically invariant quantity too otherwise it would break this symmetry of the Projective Realm.

Alternatively stated,

11.46 – the M-Set non-locality and scaling properties are the bases for the symmetry of 'relativistic invariance' of the Dualistic Projective.

I note here that whatever projective variables V1, V2, etc. might be chosen for $\Theta \sim V/h$ in order to describe the projective space it is the case that the universal dispersive scaling constant c implicit to Θ must effect them discretely and relativistically because Θ itself must uphold the scaling properties and the 'symmetry of relativistic invariance' in the Projective Realm.

Upon choosing x and t as the variables to describe the projective space all the requirements above for Θ when assembled together in the context also of our discussions hitherto of the M-Set and Quantum Mechanics lead to the general symbolic mathematical conclusion that $\Theta \sim \nu/h \sim (xp \pm Et)/h$ where 'h' is the quantum or discrete effect of the implicity 'dual' exponentiated phase variable Θ. Thence, Θ is 'quantised' too as it must be while it is also the case that because Θ is dimensionless and thence just a number Θ must be implicitly quantised. 'h' is then seen as the discrete scaling divisor of the general continuum variable 'V' of $\Theta \sim V/h$ meaning, in turn, that 'V' too is effected discretely in the Projective Realm by 'h' where 'V' has the dimensions of 'action' or angular momentum as we found in section 10.

And so it is then, too, that the coherent projected fluctuations exp (iΘ) of the Primordial Quantum Field are light-like waves; however, it is only through the miracle of projection by the simultaneous spontaneous breaking of the symmetries of self-referentiality and self-reflectivity, as we shall explain ahead, that the Quantum field appears to radiate at all because otherwise these fluctuations would be self-referentially closed loops, if you will, of exactly reflected virtual light-like waves, while it follows directly from 'Θ' above that E~h/ν, and p~h/λ if 'c' is equated 'dimensionally speaking' with a classical mechanical velocity. To see this write Θ ~ (xp ± Et)/h as p (x ± E/pt)/h so that if E/p~ν we have Θ~(x ± νt)/λ where x ± νt is a classical spatial length variable while for 'light' νλ = c, and thence from E/P = c the above relationships follow directly.

An alternative way to arrive at the relationships E ~ hν and p ~ h/λ is to exhibit the implicit scaling property of Θ by writing Θ out as Θ ~ (x / 'λ' + t/'τ') where 'λ' and 'τ' are the discrete dual divisors of the continuum variables as we discussed above which multiply to c, namely c ~ 'λ' / 'τ' ~ λ / τ ~λ ν (ν = 1 / τ). Thence Θ ~ (x + ct) / λ which when equated to Θ ~ (xp + Et) /h gives back the relationships above.

We observe too that writing Θ out in different ways reveals the implicit properties of Θ which we summarise now as the implicit universal scaling property of Θ ~ (x/'λ' + t/'τ'), the implicit universal dispersive property of Θ ~ (x + ct) / λ, and the relativistic property of Θ ~ (xp + Et)/h when xp + Et is written as the explicitly relativistically invariant scalar product of Minkovski Space '4' vectors x_μ p^μ (μ, 1–4) familiar to mathematical physics.

And finally there is the universal quantum action property of Θ, namely Θ~V/h which we shall discuss further soon, while all the properties above amalgamate to the light-like and wave-particle duality properties as well, issuing as they all do from the Virtual Computational Realm of the M-Set unto the Projective Realm of the M-Set Universe.

Now, because 'c' is actually a dispersive quantity relating to the virtual fluctuations of the M-Set and not a velocity, as might be assigned to a point particle in a classical space-time sense for example, the De Broglie relationships E~hν, P~h/λ as they have been named are in fact dispersion relationships for the extrinsic variables E and P with the result then that the M-Set leads us directly to a clear understanding of the De Broglie relationships, reflecting as they do the dispersive properties of the effects P and E dual to the space-time continuum supporting and defining them.

Looked at alternatively, the De Broglie relationships reveal the quantisation of the discrete effects of what we chose to call energy and momentum in the space-time continuum which directly relates back, in turn, to the discreteness of the effects of space and time as we have discussed it to this point.

It follows immediately too, of course, because 'c' is a universal constant dispersive property incorporating both length or space intervals and frequency or time intervals that the continuum space and time variables are necessarily bound together in '4'-vectors of a Minkovski space-time continuum in relativistic mathematical-physical parlance, and from which the current conventional mathematical formalism of relativity theory follows directly, while it is clearly the case now that

11.47 – the M-Set Universe is fully relativistically invariant.

It is also evident by now that the M-Set naturally supports the wave-quanta duality of light where the wave function or 'representation' relates to the envelope of possibilities, if you will, in the space-time continuum of the Projective Realm of the M-Set, while the quantal effects are a consequence of the Fundamental Duality of the M-Set expressing in the Projective Realm.

More generally, the wave-particle duality as expressed by the De Broglie relationships also follows naturally from the M-Set Approach, stemming back as it does to the Fundamental Duality of 'operator–operand' demanded by the Defining Property of Self-Referentiality of the M-Set.

It is also a feature of the Projective Realm of the M-Set that the wave-particle duality is apparently broken by the apparent 'collapse of the wave function' phenomenon which effectively contracts the dispersive properties of projection by the symmetries or self-measurements of the M-Set at the projective cuts that, in turn, simultaneously equates to the apparent or spontaneous breaking of the symmetry of non-locality of the M-Set to result in the apparent localisation of effects in the space-time continuum.

Stated alternatively and directly, the wave-like representations of the non-local projecting M-Set apparently collapse by symmetrically interacting or interfering with each other to create the illusion of quantal effects in a space-time continuum. In other words, one swaps, if you will, the virtual wave-like dispersive properties of the possibilities of the M-Set for a dual representation with objects or particles in a space of effecting in the Projective Realm.

Moreover, upon referring symmetry or self-measurements of the M-Set to 'observer-other' projective cut levels for which Man is the observer generating the projective cuts through measurements of 'the other' it is clear from the above that such observations will automatically produce the paradoxes of the Quantum Realm discussed thus far, while we again remind ourselves that, in all generality, all projective cuts for all of the levels of the mani-fold Projective Realm now, are, in effect, self-measurements of the M-Set, which at the human observational level we identify as the observer-other reflective self-realisations of the M-Set.

The miracle of the M-Set once again is that it is the basis for the virtual wave functions of possibilities of itself while simultaneously the Defining Property of Self-referentiality which begets the same is also responsible for mixing these possibilities symmetrically in a self-measuring interference to spontaneously create the illusion of objectively in continua which we apprehend as space-time and matter, and which simultaneous spontaneous process we shall address again in other contexts soon because of its monumental significance, especially with regard to the actual mechanisms of 'spontaneous symmetry breaking'.

The interface between the wave-like and particle-like versions of reality is explicated most succinctly by the dual angular momentum operator 'xp - px' or $x.\delta/\delta x - \delta/\delta x.x$ that contains the fundamental duality of operand–operator of the M-Set, and which through 'Θ' we know is quantised by 'h' or 'effected' discreetly by 'h'. It follows, then, that any angular momentum effect in the Projective Realm cannot be less than h, or that $xp - px \geqslant h$ giving directly the commutative relation $[x,p] \geqslant h$ known as the Heisenberg Uncertainty Principle which places the lower bound of 'h' on being able to simultaneously localise or measure dual quantities such as position (x) and momentum (p) in the Projective Realm that are, in turn, related to the Fundamental Duality of operand-operator of the M-Set expressing in this example through 'x' and $p \sim \delta/\delta x$, and similarly for the dual pair $(t, E \sim \delta/\delta t)$.

Stated alternatively, the M-Set duality and irreducible quantisation are inextricably linked and the Heisenberg Uncertainty Principle is the conventional expression of this linkage in projection.

We can now appreciate more clearly how the Fundamental Duality Property of the M-Set together with the properties and principles of projective realisation flow into the 'language' of the Projective Realm which we identify as Quantum Mechanics.

Quantum Mechanics is, in truth, a dualistically based quantised dynamic borne of self-referentiality and revealed by spontaneous symmetry breaking at the Critical Boundary of the dualistic self-projective self-realisation of the M-Set. We state therefore that

11.48 – the M-Set is the origin of Quantum Mechanics and the apparently paradoxical properties and principles of Quantum Mechanics are resolved by the M-Set because

11.49 – the M-Set is the origin of Duality.

We shall refer to this very important property again; suffice it to say here that through the defining property of the symmetry of self-referentiality the M-Set provides the context for the symmetry of duality within which the paradoxes of Duality find their re-solution or come into focus, if you will.

We have learnt to this point that the all important symmetry of duality of projective realisation is implicit to the overriding symmetry of self-referentiality, while we have also now been made aware of the fundamental operator–operand duality of the M-Set and have come face to face with the derivative dualities such as the structural–functional, wave–particle and continua–quanta dualities, all of which are so essential to the Realm of Quantum Mechanics.

Moreover, it can now be said here that through the M-Set approach the primacy or dominance of relativistic dualism has now been overthrown by the primacy of the Oneness of the Trinity of the First Cause that the M-Set is the vehicle of. This is why we now say

11.50 – the M-Set is the Holy Grail of Science (although we shall see ahead that the M-Set approach is inestimably more profound than being constrained to one 'faculty' of knowledge.)

Focusing again on the reflective fluctuations between modes or eigenstates of, for example, the Elemental State of the M-Set, and reminding ourselves here that only the fluctuations or differences between possible modes (or simply possibilities) of the M-Set emerge into the Projective Realm as the representation or wave functions at the Critical Boundary, it is more evident now why discrete effecting or quantisation must be a fundamental feature of the projective realisation of the M-Set.

We also know now that the continuum is meaningless and that there are no absolutes in the M-Set, thus only differences are defining or are 'meaningful' in projection. Moreover, it is evident that the Mutual Information or the Mutual Information Exchange implicit to the reflective fluctuations of the Primordial Quantum Field of the Elemental State, as directly expressed in the exponent 'Θ' of the representational wave functions, is necessarily quantised. To clarify this we recall that all change for the M-Set is phase change, or 'Θ' change, if you will, while upon also deferring back to the discussions of 'naïve' models of symmetry and change earlier, as well as to the discussions of the Quantum Realm we can now appreciate that 'Θ' is the implicitly quantised variable of change 'V/h'. Furthermore, the quantisation is of the so called action 'V' which in the language of classical Hamiltonian and Lagrangian dynamics relates to quantities with the dimension of angular momentum such as xp, Et, xp – px, etc.

Stated more generally, the Mutual Information Exchange of the computational dynamic of the M-Set is counted in bits of action which defines 'h' as the 'Quantum of Action' and this not only translates into the lore of Quantum mechanics that provides the underlying framework for all projective realisations but also into the classical dynamical realm through the principle of Least Action which equates with the M-Set property that

11.51 – the M-Set maximally dualises into structure and function projectively at the Critical Boundary.

To understand the latter one needs to acknowledge that the maximal redundant information reduction of the M-Set computations underpinning maximal structural–functional dualisation leads to the maximally structurally–functionally constrained dynamical outcome which, in the classical sense, is the Least Action Principle or the principle that nature is maximally parsimonious with the information available to it in order to dualise into structure and function.

Taking a different perspective for a moment, because

11.52 – the M-Set is non-dissipative,

it follows that the M-Set Universe as a whole is a perfect dynamical equilibrium which pertains at the criticality of the edge of chaos where, moreover, because of the Defining Property of Self-referentiality,

11.53 – the M-Set projective computations occur with Maximal Mutual Information co-responding with the property that

11.54 – the M-Set maximally reduces out redundant information at criticality.

Putting the above pieces together it is emerging too, as we shall further clarify when we view the M-Set as the Universal Quantum Computer in the sections ahead, that the M-Set converts the redundant information of its implicit symmetries into both meaningful structure and the mutually dually defining function at the 'criticality' of the Projective Realm, all of which amounts in turn to the property that

11.55 – the M-Set is the origin of the Universal Dynamical Principle of Least Action.

We remind ourselves again when looking at dynamical systems in a classical sense that because of the almost exact cancellation of the Trilogy of 'intendings' of the rest of the Universe in relation to the particular systemic physical aspects one is focusing upon locally to all intents and purposes it is the case that the Least Action Principle will appear to be locally true, as it is true globally for the M-Set Universe. Another way of stating this is to note that any residual effects of the renormalisations of the gravity force system over and above those relating directly to the local system of the interest to the observer are usually immeasurably infinitesimal, including in the schemes of man the observer, and we defer again to our discussion of the conservation of momentum and the billiard ball example, with the reminder once again that the properties of the M-Set translate into properties of the M-Set Universe as a whole.

It is clear now that all action in the Projective Realm is quantised although this may not be evident in macroscopic or classical dynamical settings because of the number of orders of complexification involved in the manifold projective cut of the M-Set.

We shall have more to say about the nature of quantisation in the next sections while here we need to underline several points, the first being that mathematical symbolism essentially dresses the fundamental properties and principles of the M-Set and thence is both non-fundamental and non-unique while, rather, it is the properties and principles to which the symbolism defers or about which the dressing communicates that are fundamental and unique, thence we say behind the language and symbolism of mathematics there is irreducible actuality.

The next point is that so called constraints or constants in the projective realm have values that are relative to the context of measurement. But measurement in the M-Set sense is self-measurement because

11.56 – the M-Set is symmetrical or self-measuring in the dynamical projective sense, just as it is symmetrical in the virtual self-referential sense, so the values of constants depend upon the projective perspective or point of view of the observer and because, moreover, the M-Set has no absolutes.

The size or smallness of 'h' (Planck's Constant) would then relate in the context of our observations to the orders of renormalisation of the Primordial Quantum Field by the gravity force system, equating as this does to the order of complexification or folding of the Primordial Quantum Field on itself. Relative to the 'Hydrogen' system of the Elemental State 'h' would appear of the order of '1' while to a human being it is relatively of a different order, say 1/n, where n is the number of orders of folding or complexification from the hydrogen atom to man. As a result, then, of exponentiation in projection the relative value of 'h', The Quantum of Action, might go as 1: exp(-n) (hydrogen atom: man) because the relative value is equated to the order of complexifications or folds at the Critical (exponentiated) Boundary of the Primordial Quantum Field due to the renormalisations by the Trilogy of intending of the Gravity Force System or the folding, in effect, of the Tower of Criticalities of the Elemental State of the Primordial Quantum Field in on itself. This will all become a lot clearer too when we come to explore the M-Set as a Quantum Computer. Suffice it to add here that

11.57 – the M-Set predicts relative values of quantities in the Projective Realm,

arising as they do through the symmetrical process of self-projective self-realisation of the M-Set. Paradoxically, however, the values may not appear relative when one arm of the relativism is arbitrarily normalised to '1' or to unity as with the example of 'h' above, where 'h' was arbitrarily normalised to 1 for the 'hydrogen state', the point here being that because the M-Set is self-referentially self-divisive 'relativity' is 'nested ratios' in the M-Set Realm so that arbitrarily choosing a normalisation for one level gives a number value to the other level relative to it.

'h' appears small to man because his perspective or point of view is 'n' orders of complexity beyond the order of the Hydrogen state, and in projection these orders are exponentiated by the Chaotic Computational Dynamic of the M-Set making the relative and perspective contingent value differences 'astronomical' indeed.

In all generality, the values we attribute to so called constants of the M-Set Universe reflect our perspective or point of view. In truth, the M-Set Universe is measuring itself over all orders of the manifold of itself. We are, albeit very complexly so, one of a manifold of symmetric interfaces of the M-Set's self-realisation so even constants have relative values in the M-Set Universe.

Constancy in the M-Set Universe means that 'h' and 'c', for example do not change of themselves but their values are relative.

With regard to the Gravitational Constant or to the relative value of the strength of the gravitational force system to the other fundamental force systems, which we shall discuss in the sections ahead, we note again that the Gravity System refers to the renormalisation or complexification of the Primordial Quantum Field of the Elemental State of the M-Set and hence the Quantum action associated with this force system, unlike electromagnetism that we shall identify soon with a one-phase $P(\Theta)$ system, is, topologically or physically speaking, like a surface force that ramifies by critically folding the Primordial Quantum Field.

Engaging now the physical analogies we employed in section 8 about the manifolding action of the force of gravity, in contrast to the uniphasal electromagnetic force system of the Primordial Quantum Field, the Gravitational Force System in the Projective Realm can be likened upon its first order folding to a two-phase $P(\Theta 1, \Theta 2)$ system if we also draw upon our discussions earlier about the implicit phasal degrees of freedom of physical systems. Including also our recently acquired knowledge about 'Θ', being as this is the phase variable of the representations or wave functions of the Projective Realm of the M-Set explicated by the Computational Realm of the Quantum Computer that the M-Set is, and noting moreover that the primordial phase changes or actions generated by the reflective fluctuations of the Primordial Quantum Field are simultaneously multi-plexed or manifolded by the ghost force of gravity unto projective realisation, we can then submit that the Gravitational Force System must be of the order of 'h' different in strength relative to electromagnetism on the basis of the fundamental actions of both force systems, namely $P(\Theta) \sim h$: $P(\Theta 1, \Theta 2) \sim h^2$. And indeed gravity is of this order weaker from our perspective. However, intriguingly, relative to the perspective of the Hydrogen state where $h \sim 1$ the difference in strength disappears from man's perspective while, moreover, in man's frame of reference the Gravitational Force System apparently proportionally decreases in strength relative to electromagnetism upon going up the orders of complexification in the Tower of Mass Effects. This is also a profoundly subtle property of the Gravity Force System predicted by the M-Set.

In the M-Set Universe the attraction between two distinct mass effects, for example, is related to the manifold of sub-tending between their levels of complexity or, if you will, to the manifold of possible sub-tending reflective fluctuations which underlines the in-estimably complex nature of the action of the Gravitational Force System, founded as it is on the ghost force of the Trilogy of 'intending' of the Virtual Computational Realm of the M-Set. Clearly then, the gravitational force system is a mani-fold force system of immense subtlety and 'complexity' that can only be fully comprehended by deferring to the Computational Basis of the same which the M-Set is because, as we shall see ahead,

11.58 – the M-Set is the Quantum Computer of the M-Set Universe, while it is also the case that in the M-Set perspective the relative strength of the gravity force system is subtly related to the levels of renormalisation involved at the projective cut chosen for the measurement of the same.

Another point which will also be revealed in the sections ahead is that all forces or force systems of the M-Set are the result of symmetric self-interactions or self-measurements of the M-Set taking place at projective realisation, and thence

11.59 – the M-Set projective force systems all arise from spontaneous symmetry breaking that can be visualised as the M-Set taking a cut or cross-section of itself at which self-division of unity ghost force systems appear that exactly neutralise the apparent symmetry breaking so that the M-Set actually remains perfectly symmetrical. From this we derive the dynamical notion of symetry or self-measurement of the M-Set, while it is known now that

11.60 – the M-Set is the simultaneous parallelism of all possible self-measurements, now, and it also follows from the duality of projection that

11.61 – the M-Set force systems are all dual force systems in projection.

By 'dual force system' in the M-Set Approach is meant that the functional aspects of the force system are dual to the structural aspects which in the gravity force system as we alluded to earlier amounts to gravity giving weight to itself in a dual way.

Stating the above alternatively, apparent structural projection is dual to the apparent functional forces forming the same although when there is mixing of manifold levels at more complex systemic levels of self-projection by the M-Set this is apparently not the case. Considering the example as we did earlier of the Gravity Force System, quite generally one can state that the force or function of gravity is dual to its source or structure which results in an immensely more subtle and complicated picture of extended gravitating systems in the Projective Realm than that of current classical models of gravity and which, moreover, do not need singularities such as black holes or conjectured aids such as missing mass to explain the immense and perfectly balanced dynamical systems of the cosmos which are constantly relating their story while we deny ourselves that vision.

Dual Force Systems is an area hotly pursued by current physics programs such as modern String Theories. However, without knowing the context of Duality that the M-Set is the so called 'Dual Theories' of modern physics have no basis and lead to nightmarish mathematics which, correspondingly, reflects their lost cause! In contrast, it will be revealed in the sections ahead how the M-Set quite naturally and spontaneously is the fully renormalised Quantum Field of a ten-dimensional Relativistic String, thereby extending the M-Set properties to include the property that

11.62 – the M-Set unifies Quantum Theory and String Theory completely.

The physical experiences we as an observer have of the M-Set Universe are related to symmetry or self-measurements of the M-Set at our level of observation, and so it is too that in projection the observer and observed are exactly reflected in each other because

11.63 – the M-Set projections are exactly self-reflective, even though they do not appear to be, which is a core issue we shall address when viewing the M-Set as a Quantum Computer.

Self-measurement or symmetry and self-projection are one and the same thing for the M-Set, while all projections are symmetrical in the reflective sense of observer-observed whichever cut is taken of the M-Set, and the all of it is now.

When the M-Set self-projectively self-measures it does so apparently with 'forces' whose duality reflects the duality of projection and it will be our task soon to discuss the nature of all the fundamental force systems the M-Set implicitly supports, while in the next section we shall first focus with some more hindsight on the Critical Boundary of the M-Set.

12: The Critical Boundary of the M-Set and the Phenomenon of Light

Having previously spoken about several quite general properties of the Critical Boundary of the M-Set we return here to explore in more detail properties of this aspect of the M-Set because not only is the M-Set the boundary of itself but also the Triadic Causality of the projection of the M-Set occurs at this boundary and is characterised by it, being all the result indeed of the M-Set's purely non-linear chaotic self-referential computations, while simultaneously at the Critical Boundary all manner of apparent symmetry breaking or spontaneous symmetry breaking occurs giving rise to dual force systems in such a way that the apparent functional forces exactly and dually balance the apparent structure or sources in order to realise the actuality that the M-Set remains self-referentially symmetrical because, moreover, self-referentiality can never actually be broken.

The boundary of the M-Set is called the Critical Boundary because the implicit purely chaotic wholly non-linear computations of the M-Set maximally reduce out redundant information or effect maximal structural–functional dualisation at the edge of chaos that the Critical Boundary is while the computations can be alternatively described by viewing the M-Set computations in a 'cellular automaton' sense if we identify cells with modes and note thereby that the cells compute by way of maximal mutual information transfer, relating as this does to the reflective fluctuations between the possible representations or modes of the M-Set.

Deferring to the previous discussion on Quantum mechanics, the above properties also relate to the Universal Principle of Least Action of mechanics, notably because the M-Set automatically selects the most parsimonious or efficient computations to achieve functional outcomes for structures dual to them by dint of maximal data reduction or maximal reducing out of redundant information. The M-Set in this sense is 'finding the path of least action', automatically, to provide the solution or projected outcome of itself at the Critical Boundary.

Expanding on the latter points, the Quantum of Action is the building block for the formation of structure whose dual function is played out in the space of the redundant information defining it thence, by way of duality, the Least Action Principle relates to the notion that the M-Set creates the most parsimonious structures in relation to the intended functions.

The Quantum of Action 'h' quantises the 'action function' represented generically in Θ by V = 'xp', for example, and 'the Least Action Principle' also relates to the generation of the maximal structure in the sense of data reduction or compactification, and thence to maximally meaningful function due to the maximal structure. These are also all features of a 'critical' state, physically speaking.

Now, as discussed earlier, the criticality of the M-Set computations manifests by way of the exponentiation of the representations or solutions generated by the reflective fluctuations unto dual projection, arising as they do through the implicit replications and convolutions of the wholly non-linear, non-local, self-referential M-Set that self-referentially generates the 'Quantum Fluctuations' of the Primordial Quantum Field of itself. But at the Quantum Field level these fluctuations are closed loops of spontaneous self-reflective self-interactions of the M-Set so clearly in order for self-projection to occur spontaneous or apparent symmetry breaking must also occur simultaneously as we shall discuss ahead shortly. Moreover, this simultaneous parallel chain reaction unto the symmetry breaking of projective realisation is headed by the spontaneous symmetry breaking of self-referentiality itself in relation to the ghost force of gravity with the manifestation of the Dual Gravity Force System equating to a completed renormalisation of the Primordial Quantum Field.

To state this another way

12.1 – the M-Set only projects as a fully com-pleted or fully renormalised Quantum Field relating as this does to the Tower of Mass Effects (TOME) of the Gravitational Force System which, figuratively speaking, is like stating that the Critical Boundary at projection is the completed or critically folded Primordial Quantum Field. What shows at projection are the critical com-plexifications or foldings of the Primordial Quantum Field of the Elemental State of the M-Set implicit to which is the Quantum World of the Elemental State that, in turn, the Gravity System renormalises to the surface where it is all explicated by the spontaneous symmetry breaking at the Critical Boundary of the M-Set.

As we shall soon see when exploring the elementary or subatomic ghost forces of the M-Set, the so called basic thread consisting of the Primordial Quantum Field of the Elemental State which equates with the virtual Hydrogen state is, quite literally, the doorway to the Subatomic Realm of physics while, figuratively again, the Gravity System of self-referentiality itself replicates and convolutes the Primordial Quantum Field of the virtual Hydrogen state to create the Physical Universe or the M-Set Universe by which means, in turn, the atomic and subatomic worlds are explicated, as, for example, in the Projective Realm of the turbulent dynamics of hydrogen clouds forming 'stars' wherein atomic and nuclear interactions can take place.

In other words, the Gravity Force System creates the context for the projective realisation of the Subatomic Realm at critically.

The crucial point to acknowledge here is that without completed renormalisation the realm of the Elemential State cannot be explicated and projected and, amazingly, it takes a Universe to achieve. It follows too from self-referentiality itself that

12.2 – the M-Set Universe is completed, now, in all ways, always.

The electromagnetic atomic spectra of the higher order atomic structures of the periodic table and the nuclear realm of the strongly interacting Hadronic particles of the current physics together with the weak forces of decay or transitions between Hadronic states are not singular or isolated phenomena, but rather they are phenomena which manifest in a com-pleted Universe. Thence, the critical Boundary is an unimaginably com-plexified manifold of 'the fibres of the threads' of the Primordial Quantum Field woven by the Gravity System of self-referentiality itself into the fabric or texture of the M-Set Universe.

We state that the Gravitational Force System of the Trilogy of 'intending' is the projection of the Holy Ghost Force of the Trinity or the Holy Grail that the M-Set is, which manifests through the Divinity or division of unity of the oneness of the Trinity unto the completed M-Set Universe.

Furthermore, as we shall increasingly appreciate, the distributive property of the Criticality of the Critical Boundary which is also the distributive property of the self-referentially symmetrically generated Gravity Force System is just that distributive property required at the critical Boundary to exactly neutralise or be neutralised by the dispersive property 'c' of the 'light-like' representations of the M-Set.

Alternatively stated, the dispersive property of the representations of the M-Set exactly neutralise the distributive properties of the Gravity Force System at the Critical Boundary that the M-Set is.

It is through this inestimably important conjunction created by self-referentiality itself that self-referentiality also spontaneously or apparently breaks itself about which we shall learn more soon.

We remind ourselves again that criticality ultimately is a direct consequence of self-referentiality and the pure chaos this engenders in free and spontaneous computations of the M-Set which, in turn, gives rise to the important conjunction or 'Trilogy' (the Gravity System, Pure Chaos, Criticality), as well as the property that

12.3 – the M-Set Boundary, of itself, is 'the criticality' of the fully renormalised Primordial Quantum Field of the Gravity System of self-referentiality.

As we shall discover more definitely when we understand the workings of the M-Set as the Quantum Computer ahead, the gravity force system of the M-Set Universe is an apparition of the apparent or spontaneous breaking of self-referentiality that projects the Space-Time-Matter M-Set Universe, while at the perfect criticality of the Critical Boundary one has a perfect or per-effect dynamical equilibrium between structure and function simultaneously across the M-Set manifold as a whole which, however, in localised projections appears distorted by dispersion and thermodynamical phenomena while we know that all random, probabilistic, disordered, dispersive phenomena are illusions of local perspective taking in projection.

In truth, even apparently highly disordered systems over all levels of the manifold M-set Universe reflect structural functional dualities of universal 'dynamical equilibria' because no other projection is possible, while the illusion of dissipation is the result of local perspective taking and of forgetting that there are in actuality layers upon layers, indeed manifolds of mixed layers of orders of complexification of structure dually linked to function.

Local perspective taking is always at the expense of the bigger picture and amounts in mathematical language to a projective tangent plane at some local region of the unimaginably complexified manifold of the M-Set Universe, which leads us to emphasise the very important property here that

12.4 – the M-Set properties and principles relate through the Triumvirate of the Triadic Causality at the Critical Boundary to the universal properties and principles of the M-Set Universe.

What is true for the M-Set is true for the M-Set Universe as a whole and therefore it is the case that the properties of the Critical Boundary pertain for the M-Set Universe as a whole, while we can state here that in all generality thermodynamics is the phenomenology of the Illusion of Dissipation because the M-Set is non-dissipative as we explained earlier.

The current mathematical physics frequently adopts a functional analytical perspective on the basis of the M-Set collapsed into a continuum with points. Clearly, then, through the pointilious deconstruction of the manifold one misses the immense subtleties of the universal properties and principles of the M-Set and converts the actuality of the M-Set into a meaningless mathematical analysis in a continum that only approximates projective realisation in the form of the 'classical mechanical' realm.

When perspective taking at the atomic level, however, one cannot escape into a continuum oblivion because the nature of the M-Set properties and principles dominantly assert themselves at that level or order of the M-Set Universe, and similarly at cosmological scales the M-Set provides universal properties and principles because.

12.5 – the M-Set Critical Boundary is the boundary of the Universe.

In conventional physics language the above property translates as the statement that the Critical Boundary is the Boundary Condition of the Universe and thence in the M-Set Universe

12.6 – the M-Set Universe is the boundary condition of itself, in all ways, always, now which directly implies that

12.7 – the M-Set projections are perfect solutions of the M-Set, in all ways, always, now.

We turn again to the parallelism of the process of apparent or spontaneous symmetry breaking at the Critical Boundary where the M-Set reveals the manifold layers of mixed orders of complexification of implicit dualities of structure and function of every which way the M-Set can protectively self-divide through symmetry and whereby, in turn, the spontaneous breaking of self-referentiality is associated with the appearance of the ghost force of gravity that locally generates the illusion of mass effects in a space-time continuum whose dispersive property 'c' is exactly neutralised by the distributive property of the criticality of the Boundary Condition to uphold the overriding symmetry of self-referentiality.

Incorporated or entwined, or implicated into the fabric of the fully renormalised Quantum Field that the Critical Boundary of this M-Set is, are, if you will, the fibres of the thread of the Primordial Quantum Field of the Elemental State of the M-Set from which level the symmetry of self reflectively originates to become simultaneously unimaginably complexified into the context of projection while, as we have discussed earlier, the preservation of the symmetry of the self-reflectivity directly implies that the Divinity of projective division of unity of the Oneness of the Trinity of the M-Set is exactly reflective, meaning whatever perspective or point of view, or cut one takes, 'that' is wholly reflected in 'the other'.

We know from our discussions to this point that the symmetry of self-reflectivity resides primordially at the atomic and subatomic levels of the Elemental State whose identification with the Virtual Hydrogen Atom will be further confirmed ahead with this symmetry relating directly, of course, to the reflective fluctuations of the computations of the M-Set which, without implicating the renormalising effects of the Gravity System that are simultaneously overlaid, generates the Primordal Quantum Field of the Quantum Fluctuations of the Elemental State.

Through the analogy, then, of the thread of the Primordial Quantum Field of the Elemental State of the M-Set being woven into the fabric of the Critical boundary of projection by the gravity force system of the symmetry of self-referentiality we can now see just how complexly implicated the symmetry of self-reflectivity that is begotten by self-referentiality becomes in the projective realisation, and thence how, at orders of complexity beyond the primordial level of the Elemental State the symmetry of self-reflectivity 'appears' to be hopelessly broken because of the 'mixing' of orders of complexity by the self-referential computations of the M-Set.

Alternatively, it is also evident now that when the perspective taking in projection is at the level of the 'Hydrogen state' the relatively unmixed symmetry of self-reflectively there will manifest itself more purely.

Revealed above now is the subtle interplay between the orders of com-plexity and the apparent degree of symmetry which in the physical realm relates to different scales of measurement, while noting here that because 'that' is wholly reflected in the 'other' there is no separate scale or absolute scale, nor scale breaking in the M-Set as a whole. Once again the scalelessness or property of no absolute scale of the M-Set appears to be broken because of local measurements or local perspective taking, or local cuts of the M-Set, which are associated as we now know with the apparent collapse of the wave function phenomenon.

Taking any projective cut or cross section, now, of the Critical Boundary of the M-Set we can appreciate more clearly with the knowledge of the parallelism of orders of complexification in the fabric of the Cosmos how it is that

12.8 – the M-Set projections are a mixture of all related histories.

To understand this simply look to the night sky, now, and observe that simultaneously is revealing primitive turbulent clouds of Hydrogen unto star nurseries and unto to star deaths, all in parallel too with complexification unto the human brain observing the same astride the particle accelerator pushing open the doorway to the Subatomic Realm.

All projective cuts are complete Universes.

The parallelism of all possible projective cuts of the Critical Boundary that the M-Set is, is all possible perfect solutions of the M-Set, now.

The folding of the Primordal Quantum Field of the virtual, self-referential Elemental State identified as the virtual Hydrogen State, layer upon layer unto the fully renormalised or completed Universe is physically manifested in the replication of the virtual Hydrogen state unto its turbulent dynamical elaborations into stars, elements, planets, biospheres and brains.

Naïvely put, it takes a completed Universe of 'hydrogen' to make a brain while moreover, whatever aspect you arbitrarily divide out in the projective cut 'that' is wholly reflected in the 'other', in all ways and always now.

We shall continue to return to the themes above in the sections ahead because of their inestimable significance in expressing an understanding of our Universe from the M-Set perspective, and with the pivotal place of the Critical Boundary being clearer now it is timely to focus down again on the Quantum Fluctuations of the Primordal Quantum Field whose reflective fluctuations are the home of the relatively unmixed symmetry of self-reflectivity of the M-Set.

Proceeding on the basis of the identification of the virtual Elemental State as the virtual Hydrogen State we shall turn our attendion to the modes of this state governed by the single phase P(1) connecting the Trilogy State, P(2) x 3, to the self-conceiving Trinity P(3) that parallely self-differentially self-reflects as a P(2) x P(1) state of itself with the Trilogy begotten by it, as we discussed earlier.

Taking this step by step now and identifying the P(1) 'unitary phasal' transformations' between P(1) generated modes of the the virtual Hydrogen State as the 'transitions' between 'P(2)' or 'electron' levels, or as the atomic spectra of the virtual Hydrogen State, we can thence identify the P(1) phasal system not only as the home of the symmetry of self-reflectivity of the M-Set but, moreover, we can now make the monumentally important identification that the Electromagnetic Force is the manifestation of the ghost force of the symmetry of self-reflectively at projective realisation in the same way that the Gravitational Force is the manifestation of the ghost force of self-referentiality at projective realisation.

We note again that as with identifying the Trilogy State P(2) x 3 with the Hadronic state of 3 virtual spin ½ 'quarks' we are identifying the extant P(2) State of the Trinity's self-reflective self-differentiation as the electron state of the virtual Hydrogen state of the Elemental State of the M-Set bearing in mind here the highly significant point that in the M-Set Approach the only actuality is the Virtual Realm and its properties because there is no such thing, ever, as 'a' hydrogen atom, and thence our identification of the virtual Elemental State with the virtual Hydrogen State is through the properties alone which the Defining Property of Self-referentiality determines. In this way it is the M-Set itself which has 'invented' what we have named the hydrogen Atom!

Proceeding again, we already know and will reaffirm ahead in different ways that the virtual P(1) phasal changes exponentiate at the Critical Boundary as ~exp (iΘ) or 'single phase' waves, and we know too from our discussion of the M-Set as the origin of Quantum Mechanics that exp (iΘ) expresses all the properties of light, not just visible light, and including the implicit dispersive property 'c' (spoken of in the current physics as a 'velocity') as well as the quanta of energy, ~hν, and of momentum, ~h/λ, so we are able to state that the re-flective fluctuations of the symmetry are light like and that in analogy with current physics these relate in projection to transitions between atomic eigenstates or modes of the virtual Hydrogen State which, in conventional terminology, relates to the action of the Electromagnetic Force System. Thence the electromagnetic force system in the M-Set perspective is the apparent dual physical force associated with the manifestation of the apparent or spontaneous breaking of the symmetry of self-reflectivity in the Projective Realm.

Alternatively stated, the apparent physical manifestation of the Electromagnetic Force is exactly that effect required to perfectly maintain the actual self-reflectivity of the M-Set and moreover, in keeping with our recent discussions above, this force system will be most specifically and purely evident in relationship to phenomena on the scale of atoms and their electrons or the atomic spectra.

We remind ourselves here that from the M-Set perspective because there is no such thing as a 'hydrogen atom' in projection we only see the light-like or borne on light phenomena relating to the changes or actions of the Virtual Realm of the M-Set, the Elemental State of which is unavoidably identified as the Virtual Hydrogen State because, moreover, the 'virtual phasal characteristics' of the Elemental State just happen to be those of the virtual hydrogen state.

And, that is all there is. There simply is no such thing as a hydrogen atom, or any thing for that matter.

As an important aside, which we shall explore in more detail during our foray into the subatomic forces in the near future, we note here upon identifying the virtual Hydrogen State of the M-Set as the self-reflective state of the Trilogy P(2) x 3 convoluted with P(2) x P(1) of the trinity that self-referentially and reflectively begot it that this will involve a spontaneous breaking of the internal phase space of P(2) x 3 because a particular sub-state of P(2) x 3, namely as it will turn out that of proton, is selected out from the other possible phase states of P(2) x 3 and because, moreover, only the combination of that phase state with its image in the Trinity begetting it fulfils the quantum number neutrality of the M-Set, the point we are wishing to underline here being that, in all generality

12.9 – the M-Set Universe force systems arise from the spontaneous breaking of the M-Set symmetry subserving them as it is for the Electromagnetic Force System and the spontaneous breaking of the symmetry of self-reflectivity, and as it will be shown to be for the sub-atomic force systems and the spontaneous breaking of internal phase symmetries.

It is timely now at this juncture to revisit a deep subtly of the wholly non-local, non-linear, virtual, parallel, simultaneous computations of the M-Set, namely that under the umbrella of the Defining Property of Self-referentiality itself begetting the computational properties of the

M-Set the reflective fluctuations of the virtual Elemental State amount to parallel, simultaneous replications of this state because there is no time line of the M-Set. Thence, mode 'A' of TES (the Elemental State) fluctuating to mode 'B' of TES, and similarly 'B' to 'A' is a simultaneous parallelism of mode A and mode B virtually fluctuating. This parallelism extends simultaneously to a parallelism of all possible modes of TES which complexify or fold back on themselves in a virtual parallelism of unimaginable complexity to project as the cosmos of parallel histories of interacting elements, the first point to emphasise here being that the apparent multiplicity of elements is a result of replication of TES which on cosmological scales organises projectively into entwined dual systems physically extending and manifesting unto the nurseries of stars.

The next point to emphasise in relation to the above is that the higher and higher orders of reflective fluctuations between higher order modes of the parallelisms of the M-Set computations, relating as they do to the self-referentially begotten convolutions or in-foldings of the Primordial Quantum Field of TES, co-respond, in all generality, to the action of the Gravity Force System begetting a Tower of Mass Effects that extends projectivity into the space of redundant information reduced out by these re-flective fluctuations which, in turn, belong to the so called fully renormalised Quantum Field unto the Critical Boundary of the M-Set where the virtual computations of the M-Set simultaneously support the projective realisation we call the M-Set Universe.

We shall have occasion to address these subtleties again from the perspective of the M-Set as a Quantum Computer because of their overwhelming significance, while we note here that because the primordial reflective fluctuations of the symmetry of self-reflectivity of the Elemental State of the M-Set are like the hinges on the doorway to all of the Projective Realm

12.10 – the M-Set projective realisation is 'light-like', or borne on light (not just visible light) and which property pertains, moreover, because the properties and principles of the Projective Realm are directly those of the Virtual Realm of the M-Set at the Triadic Causality of the Critical Boundary that is, in turn, the M-Set simultaneously self-projecting.

Hence

12.11 – the M-Set is never not projectivity self-realising, while it follows directly that

12.12 – the M-Set projective self-realisation is automatically relativistically dualistic, because of the dualistic light-like properties of the Primordal Reflective Fluctations, and from the vertical parallelism of the Virtual Realm of the M-Set we can say:

In the beginning there is light, now and forever.

Evidently then it is also the case that

12.13 – the M-Set projection is self-enlightenment which we shall speak of again on our journey ahead.

Continuing our discussion in a more general vein for the moment we note that in the Projective Realm the perfect duality between structure and function is not apparent because the spontaneous symmetry breaking of projective realisation at the Critical Boundary creates the illusion of force systems and dissipation. Nevertheless, when you look to complex systems such as a galaxy or the human body, for example, you realise at once that the instantaneous integrity of the system as it is perceived on a scale subtending the same must be due to the simultaneous parallelism of a manifold of unimaginably 'complexly mixed' levels of structure and function, now.

Every instant of a complex system is a perfect solution and re-solution over and over of that system, referenced in the greater whole of 'itself being reflected in the other'. And yet even as the system appears to dissipate, as in its demise, this is another possible resolution of the M-Set perfectly reflected in projection.

12.14 – The M-Set is all possible resolutions of itself.

No thing is unresolved in the M-Set Universe.

12.15 – The M-Set resolves itself now, in all ways and always.

The Divinity or the division of unity, or the splitting of the M-Set by perspective taking in projection disguises the perfect harmony of the Oneness of the M-Set, about which we shall have more to say later because this will enable us to gain deeper insights into illusions of the projective realisation to which we have frequently alluded.

The perfect dualism of structure and function of the M-Set as a whole is a 'critical phase state' or 'criticality' of the M-Set following from the properties and principles of the computational dynamic of the self-referential M-Set, thence, the M-Set Boundary is a scaleless, critical phase boundary. Without change or self-differentiation there is no projection while projection itself is change, manifesting. But change as was revealed in the section on Quantum mechanics is quantised otherwise the duality of structure and function as well as projection itself could not be realised. In projection change manifests as a force system so the quanta of change must relate in physical terms to the quanta or action of force systems or, alternatively stated, the dual force systems of projection can only cause quantum changes.

All force systems in the Projective Realm at the Critical Boundary of the M-Set are dual and quantised and appear by virtue of the spontaneous breaking of the symmetry whose representations in the M-Set form the basis for change of that force system through their reflective fluctuations which, in turn, manifest as the quantised action of the force system in the Projective Realm.

The symmetry of the force system acts in the context of the Virtual Realm of the M-Set like a potential well in which stationary vibrationary modes or eigenstates, or representations form, and between which the M-Set virtually reflectively fluctuates and simultaneously projectively self-realises by self-enlightenment.

In the context of the Virtual Realm of the M-Set and the virtual Elemental State thereof the representations or phasal modes do not have scale or are not scaled and hence the changes between the representations or eigenststes, or stationary states which are recorded in the phase 'Θ' can only be differentiated by a discrete scaleless amount, namely Θ ~n with n being a count related as we know now to the generic form $\Theta \sim xp/h$ which we observe must be a scaleless integer number. Integer due to self-referentiality ultimately because, moreover, self-referentiality requires phasal closure as occurs, for example, with stationary wave states in potential wells, so 'Θ' counts wave nodes. It follows then quite literally in the M-Set approach that only phase counts, or conversely, that counting or accounting in projection goes back to changes of the M-Set.

12.16 – The M-Set accounts for the M-Set Universe because

12.17 – the M-Set is the origin of numeracy, or counting and all that this also implies for number systems and arithmetic as well as number fields and functions which we shall expand upon more generally when discussing the M-Set as the mathematical set. We just note here in passing, however, that because the M-Set is self-referential, so then, for example, are the axioms of arithmetic while it follows directly from the Triadic Causality of the M-Set that the completion of the axioms of Arithmetic is the M-Set.

Thence, the M-Set resolves the dilemma of proof in self-referential systems as posed by the mathematician Godel because the M-set is their completion and therefore guarantees unbridled self-consistency.

12.18 – The M-Set is proof of itself, in all ways, always and now.

Moreover, a very significant property of the M-Set emerges now, namely that

12.19 – the M-Set resolves that the 'Quantum' which counts 'change' or 'action' is unique and universal.

There can be only one quantum, 'h', of change because if there were more than one size of this quantum, h1, h2 ..., this would introduce an absolute relative difference or absolute scale into the M-Set which contradicts the property of the M-Set that it has no absolute scale, deriving as this does from the defining property of self-referentiality itself. Thence 'h', the quantum of change or action is unique and universal, while it is also the case that 'h' has no absolute value because the so called value of 'h' is purely relative to the perspective or point of view taken at projective realisation, as we discussed earlier, although it is the case nevertheless that 'h' is a constant of the M-Set Universe because 'h' does not change of itself.

Once again we are observing how the M-Set can directly resolve previously imponderable questions simply as a consequence of the Defining Property of Self-referentiality.

For the 'Hydrogen State' of the Elemental State of the M-Set 'h' appears of the order of 1, while for 'man' 'h' appears relatively smaller by 1: exp(-n), n here being roughly the orders of complexification or renormalisation between atom and man.

Thus, when we refer to a constant of the Cosmos such as the so called velocity of light constant 'c' or Planck's Constant 'h', or the constant 'G' of the strength of the gravitational force system the M-Set Approach is now telling us that what constancy really means is that these quantities are evaluated relative to the projective perspective, while also being unique and universal, and unchanging of themselves.

For the M-Set that is what constancy means.

Amazingly then, the M-Set predicts the relative value of Planck's Constant 'h' for man. In other words the relative value of 'h' reflects the level of complexity at which it is perceived.

'h' is thence a relative measure of the complexity of the level of the projective perspective at which it is perceived.

The change of the reflective fluctuations is dispersed at the Critical Boundary like a phasal wave that from our discussions of the Quantum Mechanical nature of the M-Set is a light-like wave function implicit to which is the unique universal dispersive constant 'c', or so called velocity of light, and therefore the changes associated with forces in the projection are transmitted on light-like wave fronts, or on the light cone in current physics parlance, while we recall that the symmetries of scalelessnes and non-locality of the M-Set ensure the uniqueness and universality of 'c', just as we have now discussed it for Planck's Constant above.

We reiterate here that the dispersive properties of the 'representations' of the M-Set exactly neutralise the spatial distributive property of the 'critical' boundary of the M-Set computations to maintain the exact neutrality and self-referentiality of the M-Set. In other words, the apparent or spontaneous breaking of the symmetry of self-referentiality of the Critical Boundary of M-Set projection is only possible if the M-Set generates light-like representations of itself to project by. Of course it is also the case, simultaneously, that the Defining Property of Self-referentiality of the M-Set demands light-like projection at the Critical Boundary as we shall discuss further soon.

The above underlines the discovery that the M-Set Trinity gives way to Relativistic Dualism in projection, while the latter, remarkably, is also consistent with the non-local parallelism and simultaneity of the Virtual Realm of the M-Set, because all projected change is transmitted on the Light Cone where, relativistically speaking, space and time are contracted to zero. Stated differently, the actuality of the Virtual Realm of the M-Set of simultaneity, now is manifesting as relativistic causality or causality which is dispersed according to the universal constant 'c' on 'the Light Cone'.

The illusions of space and time and relativistic causality at the Critical Boundary arise from the phenomena of spontaneous symmetry breaking and the apparent collapse of the wave function, which we shall direct our attention to below with the focus also being on the symmetry of self-reflectivity and the Electromagnetic Force System.

We note, firstly, in passing that the wave functions of change of the Primordial Quantum Field of the Elemental State which are arising at the level where the symmetry of self-reflectivity is relatively explicit without the renormalisations of the Gravity Force System are in turn, and in the current physics parlance, solutions of differential equations of the free electromagnetic force field, which not only underlines the link between electromagnetism and light but also directs us to observe here that

12.20 – the M-Set only projects solutions of itself.

We know now from our discussions in the section on Quantum mechanics in relation to the Fundamental Duality of operator-operand of self-referentiality that the M-Set is the progenitor of the differential calculus of Mathematical Analysis in the spaces of realisation, so in effect

12.21 – the M-Set resolves itself through projective realisation, in all ways always and now, while it is evident that as a corollary of this property

12.22 – the M-Set is all the answers.

Alternatively stated, to ask of the M-Set in the sense of the perspective taking of projective realisation is to be directly answered.
To seek is to find in the projective realisation of the M-Set.
The quest is the Destination, now!

12.23 – The M-Set is answering as you seek the answer.

Or, quite generally now, of the Projective Realm of the M-Set Universe it can be said that as you seek so will you find.

With our focus now returning to specific aspects of the projective process we emphasise here a number of points again by recalling, firstly, that the Divinity of the M-Set is an illusion of projection because the M-Set is One, in all ways, always and thus the appearance of separateness or disunity is also an illusion of projection. Thence, in reality such a thing as a hydrogen atom, for example, does not exist. No thing exists. Projection is in the realm of that which is not.

12.24 – The M-Set is all that is, and that is One.

It is also the case that in the apparent separation of Oneness 'this' must be wholly reflected in 'that', and vice versa in order that the symmetry of reflectivity is never actually broken even though it might appear to be broken. There is also no projection from without the M-Set, about which we shall advance further discussion in the next section when we view the M-Set as a Quantum Computer while we reiterate that all projection is from within the M-Set.

There is no 'without'.

12.25 – The M-Set is the completion of the M-Set Universe, or, to state it more simply and figuratively,

12.26 – the M-Set has the M-Set Universe wrapped up.

When we talk of atomisation, like a hydrogen atom, such a thing is in truth the projective realisation of possibilities of the M-Set projecting as the representations of reflective fluctuations between the modes or possibilities or representations of the M-Set. Thence an atom is never actually separate from the whole of the M-Set. Indeed, it is never real. It is never a thing. It is forever just virtual possibilities of itself. It only appears real. But, it 'is not' in the Projective Realm.

The wave functions or representations of the M-Set are, in truth, dispersive functions of possibilities which embody the implicit duality of the M-Set and which quite literally represent all possibilities or representations of the M-Set now, but the apparent collapse of which generate the Illusion of Separateness from Oneness which Quantum Theorists refer to as the Wave Particle Duality.

At projection and thence at the Critical Boundary there is a simultaneous parallelism of subtleties of processes that we shall further elucidate here as well as into the next section where we finally pin down the nature of the illusion of the spontaneous breaking of the symmetry of self-reflectivity while we remind ourselves here that any projective cut of the M-Set involves symmetry or self-interactive self-measurement associated with both the apparent 'collapse of the wave function' phenomenon or collapse of the Dispersive Functions of Possibilities and Spontaneous Symmetry Breaking.

12.27 – The M-Set self-measurements co-respond with the apparent 'collapse of the wave function' phenomenon or to the apparent collapse of the Dispersive Functions of Possibilities that, in turn, uncovers the implicitly dual nature of relativistic causality for which the wave is the functional envelope of the quantised changes.

In co-respondence with the symmetry or the self-interactive self-measuring self-projection of the M-Set the pure duality of structure and function relating to the Fundamental Duality of operator–operand implicit to the wave functions or representations of the M-Set converts into the Spontaneous Symmetry Breaking realm of dual force systems associated with quantised entitles or sources and relativistic causality.

Evidently, then, in the simultaneous parallelisms of processes at the projective realisation the collapse of the wave function phenomenon is also a Spontaneous Symmetry Breaking because a choice in the virtual space of possibilities generated by the symmetry or projective self-measurement of the M-Set is being made, so generally it is the case that

12.28 – the M-Set apparently or spontaneously breaks its symmetries in correspondence with the apparent collapse of its representations or wave functions, while it is also the case now that

12.29 – the M-Set sym-metry or self-projective self-measurements correspondingly spontaneously break the symmetries of the M-Set.

We observe now that the properties 12.27 to 12.29 give us the Trilogy of (symmetry or self-measurement, apparent 'collapse of wave function' phenomenon, Spontaneous Symmetry Breaking), with each aspect of the Trilogy simultaneously co-responding with the other two aspects while, simultaneously, all aspects co-respond as one. This is the nature of all Trilogies, connecting, as they do, the Oneness of the M-Set with the Relativistic Dualism of the Projective Realm.

With regard to the projective realisation of the M-Set at the Critical Boundary the M-Set can also be visualised as a self-referentially closed potential well supporting all possible reflective or stationary states of itself that projectively interfere in all possible ways, with the interference being considered synonymous with the self-measuring collapse of the wave function phenomenon to generate spontaneous or apparent symmetry breaking.

The pure dualities implicit to the wave functions or 'representations' of the M-Set parallel the dual force system realisations of the Triadic Causality at the Critical Boundary where, in turn, the Oneness of the M-Set is explicitly dualising through the Trinity Principle, while both the Oneness and the Duality of the Triadic Causality of the M-Set are simultaneous, and both are as One because the M-Set is all there is.

We know now that the distributive property of projected phenomena in space and time must be fractal or go as ~ '1/ι x 1/f' at the Critical Boundary of the M-Set because of the purely chaotic nature of the M-Set computations, although from the enormous mixing of orders of complexity pure strains of criticality can become complicated or folded into each other in seemingly unraveallable ways, and thence only in relatively pure or primordial, or less complex systems, is critical fracticality most evident such as on Cosmic scales or in predominantly gravitating systems as we explained earlier.

We also know that the dispersive property of the wave functions or representations of the M-Set is implicit to a unique and universal constant 'c' ~ λν which is called the velocity of light' and which property exactly neutralises the apparent physical, critical, fractal distributive property of the projective realisation. That is, symbolically, $\lambda v/\iota.f \sim 1$ where λ and v do not just refer to visible light scales but to relative scaling of space-time phenomena.

A correct interpretation of 'c' is thence considerably more profound than as a velocity because 'c' is inextricably and simultaneously linked to the fractal scaling laws of the M-Set Universe over all orders and levels of complexity.

The relative characteristic value of 'c' depends upon the relative scale of the observer. To a galaxy c appears relatively much smaller than for an atom. For man it has a characteristic value too in the same way that Planck's Constant has relative values while being universally unique.

The symmetry of reflectively will evidently be most pure at the level of the representations or the wave functions' of the Elemental State of the M-Set that, in turn, the gravity system of self-referentiality simultaneously weaves into a critical Tower of Mass Effects which involves unimaginably complicated mixing of the Primordial Quantum Field or the entwining of the thread into the fabric of the fully renormalised Quantum Field. In effect, then, the virtual hydrogen atom whose identification with the Elemental State will be elaborated upon soon is the portal through which we can view aspects of the M-Set in an unravelled way, including the Electromagnetic Force System and the symmetry of self-reflectivity as we are about to explore further.

Remarkably, as we explained earlier, because of the simultaneously linked properties of the criticality of the M-Set Boundary and the light-like wave properties of the representations of the M-Set, the M-Set spontaneously linearises in projection which is expressed mathematically as the Hilbert Space of States. The M-Set, in turn, is like a self-referentially bound potential well or pond of reflected interference patterns of the virtual waves of all possible modes of itself and in which pond or well every wave mode, regardless of the teeming complexity of the interference patterns, maintains its integrity and independence because there is no dissipation in the Virtual Realm of the M-Set, from whence it emerges too that the spontaneously generated representations or wave functions of the reflective fluctuations between eigenstates or modes of the M-Set act as independent' degrees of freedom in the abstract linear space of states called the Hilbert Space of States.

The significance now of the miracle of projective linearisation of the wholly non-linear M-Set is that mode by mode, regardless of the unimaginable complications of the mixing of the symmetry of reflectivity by the symmetry of self-referentiality, the symmetry of self-reflectivity is implicity upheld, and which property of the M-Set we are about to address more specifically in the next section when we look again to this highly important property from a computational perspective to further pin down the illusory magic of the M-Set.

With the above in mind we focus again on the wave functions or representations of the reflective fluctuations between any two representations or possible modes of the virtual hydrogen atom or the Elemental State that forms part of the Elementary Spectrum of the M-Set, and which equates in projection to the Hydrogen Atom Spectrum. Now, every independent representation or wave function of the Hillbert Space of States must have a perfect reflection or reflected wave function in the M-Set Universe because the M-Set is perfectly neutral, which translates, mathematically speaking, to the Quantum Mechanical property that in order for the wave function ψ or Dispersive Function of Possibilities to represent physical projection it must have a reflection such that $\psi\psi^R \sim 1$, or is normalised to 1. In the language of Quantum Mechanics and the current physics this relates directly to the so called Hermitean property of physical wave functions, $\psi\psi^+ \sim 1$.

Taking the generic wave function ψ as exp ($\pm$ iΘ) or (expi(sign) Θ, sign = $\pm$ 1) we now observe the elementary property from the Quantum Mechanical structure of $\Theta \sim (xp \pm Et)/h$ that if we change the signs of sign, 'x', and 't' simultaneously the wave function remains unchanged which leads to a profound result of fundamental physics, namely that the Projective Realm is invariant to the simultaneous triple symmetry operation of parity reversal, P x ~ - x, time reversal T t ~ - t, and charge conjugation C sign ~ - sign provided we also make the profound identification in the M-Set Approach that the bare 'charge' or unrenormalised charge is simply the signature 'sign' $\pm$ 1, thence not only rendering Quantum mechanics and Quantum Field Theory automatically 'PCT' invariant, as is true for the physical universe, but also revealing that charge, deep down, is actually the signature of the symmetry of reflectivity.

The significance of this identification for the virtual, primordial, un-renormalised 'charge' will become evident in the sections ahead where the fundamental physical force systems will be explored together suffice it to say here that charge gains its physical presence through the renormalisation of the gravity force system at the Critical Boundary where the M-Set projectively realises with Spontaneous Symmetry Breaking and the emergence of dual force systems, including that of electromagnetism. At the level of the Primordial Quantum Field charge has a reflection of itself (the Looking Glass Universe) in the conjugate or reflected primordial wave function or representations of the Elemental State with the reflected charge residing in a parity and time reversed projective realm.

Clearly now, the complexification by the renormalisations of the gravity Force System of self-referentiality will complicate the Primordial Quantum Field and promote the creation of the illusion of the apparent breaking of pure reflectivity which we shall explore further in the next section. This will also simultaneously give body to the bare charge while we add here that at the Critical Boundary because the representations ψ are PCT invariant, namely PCT $\psi = \psi$, it is a purely arbitrary assignation or a matter of words whether we live in a Universe of charged particles with time t going in one direction and handedness that is left or a Universe of antiparticles with opposite charge with time t going in the other direction and right handedness with the Parity operation 'P' quite generally relating spatial orientations that are mirror reflections of each other so that the symbolically trivial example of P $\psi \sim - \psi$ extends to torque and spin or handedness that become mirror reversed under parity reversal.

The two PCT choices are identical, however, once an assignation is made in projection there appears to be an asymmetry, but this is an illusion of projection created by perspective taking' or one's point of view in relation to the complexifying renormalisations of the implicitly reflectively symmetric primordial representations as we shall elaborate upon further in the next section.

At this juncture we shall draw together a number of outcomes of the M-Set approach in relation to the Electromagnetic Force System by noting, firstly, that it is natural to locate this force system with the apparent or spontaneous breaking of the symmetry of self-reflectivity because the representations or wave functions of the Primordial Quantum Field are light-like waves as well as solutions of the free or unrenormalised Electromagnetic Force Field Equations in the current physics while, moreover, the phenomenon of bare or virtual charge which, in the M-Set perspective, is the signature of reflectivity itself is always renormalised into a carrier mass effect, or a particle such as an electron or proton in physical realisation, and this is the work of the Gravity Force System.

To put this another way, at the 'conception' of the M-Set Universe which is the Trinity Principle the primordial vehicle of charge separation, namely the virtual unrenormalised 'hydrogen State' or the Elemental State is established. The Elemental State is like the seed of charge separation which is simultaneously multiplied by the M-Set computations unto all levels of complexification of the M-Set manifold Universe.

It follows too in the Projective Realisation of the M-Set that is the M-Set Universe that the most elemental level at which the apparent or spontaneous breaking of self-reflectivity will manifest is the atomic structure and function of the renormalised or projected Hydrogen State at the Critical Boundary of the M-Set.

In other words, the immense complexifications of the renormalisations of the Gravity Force System of the symmetry of self-referentiality locks in the apparent charge separation of the conception that is the Trinity Principle about which we shall also have more to say in the next section because of its inestimable significance in the illusion of spontaneous creation, now.

In effect also, the renormalisations of the Gravity System dress the bare charge to create the mass effects of particles which in the context of the manifold of the M-Set Universe appear to interact objectively in the field of space-time, while we know now that in truth no thing is separate and that these illusions are the result of the 'collapse of the wave function' phenomenon of perfectly self-referential and self-reflectively symmetric computationally complexified reflective fluctuations interfering in all possible ways in the deep well of possibilities of the M-Set which, in turn, is projectively realised, mathematico-physically speaking, as the Linear Superposition of representations that we call the Hilbert Space of States of the Projective Realm of the M-Set.

To conclude this section we shall provide an example of how the M-Set can answer seemingly imponderable questions the question we ask here being, how is visible light in fact visible? Or, how is it that the Hydrogen Atom Spectrum or the 'Atomic' Spectrum is the visible range?

Drawing upon what we now know to this point, we observe that at the conception of the M-Set Universe, which is the Trinity Principle, there is light. Now viewing the most primordial level as the beginning, while recalling that this level is in parallel with all other levels of the manifold of the M-Set Universe, because of the simultaneous parallelism of the virtual computational realm of the M-Set we can say 'In the beginning there is light.'

Now, we know also that for the M-Set there is no space and there is no time, and that the M-Set is that. Thence for the M-Set there is no question of where or when, or why. There simply is no place of where, no time of when, and no why of that which simply is. There is only one question one can ask of the M-Set, and that is 'how'.

12.30 – The M-Set only answers 'how'.

The answers to all other questions are relative to ones perspective or point of view in the Projective Realm. Who is also relative. We are all related! And what is, is. What is not, is not.

How is it then that the light of the inception is the 'light of our vision'?

The M-Set answers thus; as light is given so it is received.

And so it is in our mind's eye that the light which is given is received at the atomic level.

And so it is for the M-Set Universe that as it is given so will it be received because the M-Set is one, in all ways, always now.

In the next section we shall continue to explore the symmetry of reflectivity, however, from what we now know we can say

12.31 – the M-Set Universe is a looking glass universe because, regardless of the perspective or point of view, 'this' is reflected in 'that', and 'which' is which is relative too because both are One.

Reality is a reflection of itself.

The illusion of reality is the reflection of One, in all ways and always, now.

13: The M-Set as Quantum Computer and the Symmetry of Reflectivity

Remarkably,

13.1 – the M-Set is the Universal Quantum Computer being as it also is the origin of Quantum Mechanics and elaborating its computations from the reflective fluctuations of the Elemental State that, when fully renormalised, generates a Quantum Field of such unimaginable computational power it underpins the projective realisations manifesting as the 'observable' manifold we call the M-Set Universe.

There are a number of features of the M-Set 'Quantum Computer' we wish to assemble here that will consolidate some of the properties and principles illuminated to this stage, while we shall also secure a point of reference for further discussions.

It is already known now that

13.2 – the M-Set Elemental State is irreducible and unique providing as it does the primitive modes of the primordial reflective fluctuations which generate in turn the Primordial Quantum Field whose renormalisations manifest in the projective realm as the action of the Gravity Force System of the Tower of Mass Effects.

The Elemental State is thence the fundamental unit of all of the M-Set computations, just as it is observed in the M-Set Universe that the Hydrogen State is the fundamental irreducible ingredient of cosmology as well as the doorway to the Subatomic Realm prised open by the infernos of turbulent hydrogen unto stars which 'maketh man and machine'.

As we shall explore further, the Fundamental Computational Unit is conceived by the Trinity Principle of the M-Set and is therefore at the beginning, now, in all ways, and always.

We also know from the Defining Property of Self-Referentiality of the M-Set that by nature

13.3 – the M-Set self-computations are purely non-linear and thence purely chaotic which implies that the dynamic of these computations is one of pure mixing or convolution, from all of which it follows directly that

13.4 – the M-Set computations are cyclical because all computations are folded or kneaded back upon themselves over all orders of renormalisation or self-referentially bound complexification, leading in turn to the crucial property that

13.5 – the M-Set computations are free of dissipation meaning that all variations or perturbations implicit to the computations are folded back into the whole of it.

It is also the case then that

13.6 – the M-Set Universe is non-dissipative as a whole while locally in projection it will appear as if dissipation is a property of the M-Set Universe. But it is not.

13.7 – The M-Set Universe as a whole is a perfect balance of the dispersive and the distributive properties of itself. It follows generally from the above too that

13.8 – the M-Set is revolutionary because it refolds into itself with the effect that the parallel, virtual computations generate a simultaneous cyclical waxing and waning between involution and evolution or in-folding and un-furling in the convolutionary computational mixing of the reflective fluctuations.

We know too that

13.9 – the M-Set computations are spontaneous because

13.10 – the M-Set is perfectly free, expressing as it does in the essential vitality of the M-Set Universe that we call the life force which manifests with the perfect (per-effect) relativistic dualism and the perfect dynamical equilibrium of the Projective Realm we call the M-Set Universe.

From the perspective now of the M-Set as a Quantum Computer we also know that in order for the wholly non-linear, purely chaotic, mixing, convolutionary computational dynamic to project it must be able to spontaneously linearise itself or form a linear representation of itself, at the heart of which resides the symmetry of reflectivity that underpins the generation of the reflective fluctuations from the representations of the primitive modes and their complexifications to enable a projective realisation by way of a 'linear representation' called the Hilbert Space of States.

Figuratively speaking, the self-referentiality of the M-Set can be regarded like a self-bounded pond in which self-referentially bounded reflective fluctuations spontaneously occur that originate, in the M-Set perspective, from reflections between the primitive modes or representations of the primordial self-differentiation and self-conception we call the Trinity Principle which, in turn, spontaneously creates the 'spirit water' Hydrogen or, if you will, the Holy Water of the Holy Grail that the M-Set is, and of which primordial solution the virtual, parallel, simultaneous ripples or reflective fluctuations simply are.

Continuing the 'pond' analogy, the Hilbert Space of States corresponds to a linear superposition of the individual virtual fluctuations because in the 'pond', while all these waves interfere in every possible way they never dissipate and thereby maintain their identity or independence which gives rise to the mathematical notion of a space spanned by the independent waves or representations, where these are regarded like orthogonal axes in an abstract space. Thence, we discover the famous Linear Superposition Principle of Quantum Mechanics which we now understand as an essential property of the Projective Realm of the M-Set.

We know also because of the virtual, non-linear, parallel, combinatorial, convolutionary nature of the computations and the unitary phasal nature of any change as well as the critical property of the boundary of the M-Set that the representations must be wave functions of the form exp $(i\Theta)$ which we have already spoken about at some length, and which functional forms also uniquely mathematically enable a spontaneous projective linearisation of the M-Set manifesting as the famous Linear Superposition Principle.

The virtual waves of the self-referential pond relate to the Imaginary Part of exp $(i\Theta)$, or Im exp $(i\Theta)$ in standard notation where Im exp $(i\Theta)$ ~ sine (Θ) is the 'sinusoidal function' of Θ, embodying as it does in Θ the quantised action of the change implicit to the reflective fluctuations that define the same, and which in turn exponentiate the change in the Representative Wave Functions at the Critical Boundary of projection of the M-Set.

Change, as we learnt earlier, is related to the Mutual Information Exchange or Symmetrical Information Exchange (mnemonically SINE) of the re-flective fluctuations between representations or modes of the M-Set. We learnt also that the Mutual Information is maximal for the M-Set computations which has many direct consequences for the M-Set as a Quantum Computer, not the least of which is that

13.11 – the M-Set maximally reduces out redundant information. Therefore

13.12 – the M-Set maximally computes because 13.11 implies 'maximal data compression', if you will, as well as maximal structural–functional dualisation which miraculously uniquely pertains at criticality or 'on the edge of chaos' of the purely chaotic computations of the M-Set.

Criticality and maximal structural–functional dualisation are one and the same thing in the M-Set perspective.

There is no more efficient or complete computation than the computations of the Quantum Computer that the M-Set is.

13.13 – The M-Set is the upper bound of computational power while it is clear

13.14 – the M-Set spontaneously computes, now, in all ways, always, and that

13.15 – the M-Set is the Computational Support of the M-Set Universe.

Expanding our discussion again we note that the M-Set computations are a virtual parallel mixture of all orders of self-compounding reflective fluctuations, which projectively translates as a linear superposition of representations or wave functions of all possible modular reflections arising between the primitive modes of the Elemental State, all the way up to the unimaginably complexified modular states of the renormalised levels of the Primordial Quantum Field. The completed linear space of representations is referred to in the language of Quantum Mechanics and Quantum Field Theory as the Hilbert Space of States, and in the self-referential pond analogy it is seen as an automatic consequence of the M-Set computations.

Stated alternatively, self-referentially bound self-reflectivity implies a projected state of a linear superposition of wave functions representing the reflective fluctuations of the representations of all possible renormalisations of the Primordial Quantum Field of the M-Set.

This is also why we say that the M-Set self-referentially, self-reflectively, self-representationally, self-realises self-projectively.

Ascending to higher and higher levels of complexification and thence to higher levels of modularisations it will become evident soon because of the symmetry of self-reflectivity that, projectively speaking, 'this' is reflected in the 'other' over all levels of projective realisation. Thence it is the case that for all projective cuts of the M-Set.

13.16 – The M-Set is a mirror image of itself, projectively. Invoking the Primordial or Fundamental duality of operator–operand of self-referentiality, every thing that is apparently projected by the M-Set is reflected structurally and functionally in the 'other'.

13.17 – The M-Set spontaneous creation is in the image of itself in the M-Set Universe, in all ways, always and now.

We are thence all spontaneously created in the image of the Oneness of the M-Set.

Alternatively, upon using the language of the symmetry of reflectivity again we can say

13.18 – the M-Set is self reflective or folds back into itself, computationally speaking, while for every projective cut it is evidently the case that

13.19 – The M-Set Universe is a 'reflection' of itself, in all ways always and now.

The self-referentially bound symmetry of self-reflectivity begotten by the overriding symmetry of self-referentiality clearly relates to many pivotal properties of the M-Set, including the reflective nature of projection, the 'quantisation' of change implicit to the 'self-referential pond' of reflective fluctuations of the M-Set, and the cyclical or revolutionary nature of change in the projective realm which in turn underpins the dictum that history repeats itself, but with the understanding now that the latter property extends to every historical level of the projective realm from the life cycles of cells to economic cycles and on out to the cycles of the births and deaths of stars all nested in the non-local, virtual parallelism of the reflective fluctuations of the renormalisations of the Primordial Quantum Field of the M-Set Quantum Computer, now.

Focusing again on the Trinity Principle we make the observation here that the self-conception of the Trinity is the primordial self-differentiationof the M-Set, associated as it is to the symmetry of reflectivity, while it is the direct consequence of the overriding symmetry of self-referentiality that the reflective fluctuations are replicated or summed, or, in mathematical language, are integrated.

It is a profoundly deep insight now that the symmetry of self-reflectivity is the symmetry of self-differention and the symmetry of self-referentiality is the symmetry of self-integration in the computational realm of the Quantum Computer of the M-Set, with all the renormalisations of the Gravity Force System of the Projective Realm corresponding to the integrative computational functions while simultaneously the differential functions, including those of projection of the Tangent Plane of the linear Hilbert Space of States, relate to the symmetry of reflectivity of the Electromagnetic Force System manifesting, as we now know, in the relativistic causality and relativistic dualism of the Projective Realm of the M-Set Universe.

In effect, then, the projective realisation is an 'enlightenment' of the integrative functions of self-referentiality by the primordial differentiation of the symmetry of self-reflectivity of the Trinity Principle.

Stated alternatively, the profound insight above is informing us that the Projective Realm directly derives its properties from the differential nature of the symmetry of self-reflectivity while, as we shall elaborate upon soon, the mass effects or apparent substantiality of the Projective Realm is due to the integrative role of the symmetry of self-referentiality in the Computational Realm of the M-Set that, projectively, is enlightened by the self-conception of the Trinity Principle.

All projection is enlightenment. Spontaneous creation is self-enlightenment. We live in a 'world' of virtual reality.

13.20 – The M-Set Quantum Computer self-projects by self-enlightenment and thence

13.21 – the M-Set Universe is the self-enlightenment of the M-Set.

It can be said then that we are all enlightened beings and that reality is the illusion of enlightenment.

Moreover, it can also be said that the Divinity is the enlightenment of the Trinity that is One and thence that we are all Divine Beings. The all of it is Divine.

Continuing our general discussion we have now arrived at the profound computational property of the M-Set, namely that

13.22 – the M-Set Quantum Computer simultaneously integrates and differentiates in a perfectly dual way across all orders of complexification.

The highly significant global implication of property 13.22 is that all solutions of the M-Set computations are light-like or wave-like, or indeed exponential representations (in mathematical functional terms) in the Projective Realm because the exponential functional representation is the only functional representation which remains the same simultaneously for integration and differentiation, namely integral 'S' exp (iΘ) ~'δ/δ' exp(iΘ).

Property 13.22 of the M-Set provides further confirmation of the exponential 'functional nature' of the projective representations of the M-Set computations in the Projective Realm, reaffirming once again that all M-Set projecting is borne in 'light-like solutions' or on the Light Cone.

We can begin to see that it is the integrative 'complexifications' of the mixing dynamic of the ghost gravity force system of the M-Set Quantum Computer which creates the illusion of substantiality of projective realisation, enlightened simultaneously as it is by the self-differentiation of the Divinity of the Trinity Principle of the Oneness of the M-Set that the projective cut of projective realisation is.

In effect, the 'apparent substantiality' of the Tower of Mass Effects (TOME) relates to unimaginably complicated interference patterns of replicated convolutions of the primordial reflective fluctuations in the 'self-referential pond' represented by the Hilbert Space of States that in turn all projects locally with the apparent collapse of the wave function while,simultaneously, the M-Set symmetry or self-measurement spontaneously creates the illusion of the self-enlightenment of the Divinity or 'division of unity' of the Trinity that is One for whichever projective cut is chosen.

The profound property of 13.22 is also the basis for now stating that

13.23 – the M-Set unifies Special Relativity with General Relativity, completely through the Defining Property of Self-referentiality itself, while such a complete union is clearly required for critical, dual projective realisation.

Broadly speaking, the gravitational force of the symmetry of self-referentiality signifies the integrative functions of the M-Set Quantum Computer which are differentiated by or atomised by the Electromagnetic Force System of the symmetry of self-reflectivity and deep to which, as we shall explore in the next sections, are the subatomic force systems.

From the M-Set perspective the unification of the ghost force systems of gravity and electromagnetism relating respectively to the General and Special Theories of Relativity is ultimately wholly because of the Defining Property of Self-referentiality of the M-Set under whose umbrella the

Sub-atomic 'strong', 'nuclear' and 'weak' force systems of high energy physics must also reside which, in turn, completes the complement of the fully unified fundamental physical force systems of the M-Set Universe of the M-Set that is One.

Turning our attention again to the purely chaotic computations of the M-Set Quantum Computer we highlight here that despite the enormity of the wholly non-linear chaotic mixing of the complexifications of the self-referentially bound self-reflective computations these miraculously line up as light-like representations in the linear Hilbert Space of States. Implicit to this miracle of linearisation are a number of deep subtleties the first of which is the spontaneous breaking of the symmetry of self-reflectivity.

We know from the symmetry of self-reflectivity that

13.24 – the M-Set is a perfect reflection of itself across all orders of renormalisation of the reflective fluctuations, and as also demanded by the symmetry of self-referentiality itself so how, one might ask, does a perfectly reflectively symmetric state give way to the evident asymmetry of the 'this' and the 'that' of our material universe? The answer, of course, is the simultaneous mixing or the self-referential integrative computational functions of the M-Set Quantum Computer that jumbles up all the orders of the perfectly reflectively symmetric fluctuations of the perfectly self-referential M-Set to apparently break the symmetry of self-reflectivity in the Projective Realm which is also why we said earlier that in projection the 'this' is perfectly reflected in the 'other', in all ways, always, now.

Stated alternatively, the integrative functions of the computations of the M-Set Quantum Computer apparently or spontaneously break the symmetry of self-reflectivity, while, in physical terms, the Gravitational Force System dresses the bare charge with mass and a perfectly reflective Universe of 'this' reflected in the 'other' is spontaneously created.

The symmetry of self-reflectivity is never actually broken. It only appears to be broken in the Projective Realm.

Simultaneously, the differential functions of the computations of the M-Set Quantum Computer apparently or spontaneously break the symmetry of self-referentiality because it is the simultaneity of the differential and integrative functions of the M-Set Quantum Computer that generates the light-like representations of the M-Set whose dispersive property 'c' is exactly neutralised by the fractal distribution '1/ι.f' of the purely chaotic computational dynamic of the M-Set Quantum Computer.

13.25 – The M-Set is exactly neutral, in all ways, always and now.

This is the miracle of the spontaneous breaking of the symmetry of the self-referentially defining the M-Set.

Consequently

13.26 – the M-Set Universe is perfectly reflective of itself, in all ways, always and now, regardless, then, of which projective cut, perspective, point of view, or position of reference is taken, 'this' is perfectly reflected in the 'other' in the unimaginably immense parallelism of the manifold M-Set Universe.

The next question of clarification one might ask is: how do the unimaginably mixed up reflective fluctuations of the M-Set Quantum Computer miraculously line up to form the linear Hilbert Space of States? And the answer to this, of course, is because

13.27 – the M-Set simultaneously integrates and differentiates itself in an exactly 'dual' way as demanded by self-referentiality that uniquely generates exponential or sinusoidal wave function representations of the reflective fluctuations as we discussed earlier, while in the non-dissipative self-referential pond analogy the representations form the independent axes of the abstract linear Hilbert Space of States, embodying directly as this does the Linear Superposition Principle of the Interference Patterns of the Self-referential Potential Well of the spontaneous reflective fluctuations between possibilities of the deep well that the M-Set is.

There is a perfect dualism between the spontaneous breaking of self-reflectivity and the spontaneous breaking of self-referentiality because at the Divinity or self-projective self-enlightenment of the M-Set there appears to be a division of unity that, in turn, appears to break self-referentiality, however, it is the pure or actually unbroken symmetry of self-reflectivity which exactly neutralises the apparent separation of the 'this' from the 'that' in the Projective Realm, as manifesting physically in the perfect balance between the dispersive and distributive properties of the M-Set Universe as a whole.

By way of clarification we can state that the ghost force of gravity of the symmetry of self-referentiality that manifests in the critical distributive property of the M-Set Universe is exactly neutralised by the 'dispersive' property of the ghost force of electromagnetism of the symmetry of reflectivity with these symmetries simultaneously and dually spontaneously breaking each other to generate the Divine spontaneous self-enlightment of the projective realisation of the M-Set we call the M-Set Universe.

The dualism of self-referentiality and self-reflectivity not only performs the Miracle of linearisation of the projective realisation of the M-Set Quantum Computer but it also performs the miracle of generating the Critical Boundary, the distributive properties of which in the sense of TOME is exactly neutralised by the dispersive property 'c' of the light which reflectively bears the projection or spontaneous creation of the M-Set Universe. This is the act of the spontaneous breaking of the symmetry of self-referentiality itself generating thereby the greatest illusion of them all, namely the 'Illusion of the Disunity' of that which is One.

Things seem to be apart.

But they are not.

They are One, in all ways, always and now.

And so it is that

13.28 – the M-Set Quantum Computer generates solutions of itself through self-referential, self-reflective, self-representational, self-projective, self-realisation.

Miraculously, the perfect self-reflective self-referential M-Set Quantum Computer can spontaneously generate solutions of itself because the simultaneous dualism of the computational processes of self-integration and self-differentiation occurring in the context of the mixing dynamic of self-referentiality enables the perfectly reflective states of the M-Set to become all mixed up, while maintaining the symmetry of reflectivity perfectly unto both the Critical Boundary and the Projective Plane of the Hilbert Space of States.

Evidently

13.29 – the M-Set creates the illusion of reality through the spontaneous breaking of its defining symmetry of self-referentiality and we say that

13.30 – the M-Set spontaneously projects, in all ways, always and now.

Moreover, at the Triadic Causality of the Critical Boundary that the M-Set is, the M-Set is simultaneously an unimaginably immense parallelism of mani-fold Universes begotten because

13.31 – the M-Set 'self-divides' by spontaneously dually breaking the symmetries of self-referentiality and self-reflectivity in all ways, always, and now.

The Cosmic 'cake' is all possible slices of itself, now.

The Cosmic cake is never consumed no matter how many different ways it is experienced by the different cuts of itself.

In the beginning it is the symmetry of self-referentiality of the M-Set that makes the ingredients, mixes the cake and serves it up!

That's Magic.

Hence

13.32 – the M-Set is com-pletely self-consistent because the M-Set consists of itself, and so it is that

13.33 – the M-Set is the Alpha and the Omega of the M-Set Universe because there is no other.

We now take a different tack towards illuminating the spontaneous creativity of the M-Set by focusing firstly on the modes of the Elemental State, which we note here are never actually projected, rather, only the transitions or changes, or differences between these possible modes are ever projected through the reflective fluctuations between them as discussed in the section on Quantum Mechanics for example.

Moreover, in the context of the Virtual Computational Realm of the M-Set Quantum Computer the reflective fluctuations correspond to unitary phasal transformations in a 'mathematical' sense because these leave the M-Set invariant or 'unchanged'.

Alternatively stated, we are making the observation generally that transformations corresponding to symmetry operations relating to the implicit symmetries of the M-Set must be unitary operations, or like unitary phasal operations in a mathematical representational sense because only these operations leave the Unitary M-Set 'unchanged'.

Now, each possible Hilbert Space of States corresponds to the representative space of a possible projective cut of the M-Set of which there clearly is an indefinite number, depending upon ones' perspective or point of view, while it follows directly from the observation above that every possible Hilbert Space of States or Projective Plane is connected to the others by unitary symmetry transformations of the M-Set.

This is a remarkable property of the Projective Realm of the M-Set which has profound consequences, while we note here that each possible Hilbert Space of States is a possible wave function of the M-Set Universe which translates into the property that

13.34 – the M-Set re-presents the M-Set Universe, in all ways, always, now whereupon

13.35 – the M-Set Universe is a parallelism of all possible representations of the M-Set, now.

The above properties of the Projective Realm also have profound consequences for Quantum Dynamics and Quantum Field Theories of theoretical physics with the observation here, firstly, that because of the Fundamental Duality of operator–operand of the M-Set the so called matrical operator or 'matrical mechanical' and the Wave Function approaches to the formulation of Modern Quantum Physics are exactly dual mathematical descriptions of physical reality.

Moreover, from what we now know about the M-Set it is clear that the Matrix Mechanical approach, involving as it does the unitary matrical transformation tools of Mathematics in relation to the reflective fluctuations of the M-Set Quantum Computer, is most efficacious for describing phenomena down at quantal levels, as for example when exploring the properties of the atomic spectra of the Hydrogen state or Elemental States while the description at the level of the wave functions or representations of the M-Set Projective Realm, involving as it does the mathematical notions of phase transformations, is most efficaciously applied to multi-system analysis or multi-particle systems and all of which is familiar to practising Theoretical Physicists that we shall also have occasion to refer to again when we explore the subatomic force systems implicit to the Elemental State of the M-Set.

I observe here that all possible unitary transformations of the M-Set correspond to all possible ways the M-Set can self-differentiate, which, in the context of the Projective Realm, corresponds to changes in the projective cut of the M-Set that, in turn, spontaneously creates the illusion of motion in space-time.

A tangible analogy of this might be the images projected on a TV screen of a video game that provide animation of the program of the game which is on the disc. The program can be viewed as a parallelism of all possibilities of the game, now, in concert with the M-Set being the 'deep well of possibilities' of its own projective realisations, now, while the frame to frame changes creating the illusions of motion and of space and time on the screen relate, in the M-Set approach, to changes in the projective cuts which are connected by self-differentiations of the M-Set. Thus, in a sense, the projective cuts move through the field of possibilities every which way to generate an animated realisation we call the M-Set Universe bearing in mind now that this Universe refers to parallelisms of manifolds of possibilities of itself.

We now know, however, that if self-differentiation was all the M-Set were doing there would be no projection because the M-Set Universe would self-reflectively disappear without the simultaneous dual self-integration of the symmetry of self-referentiality which mixes up the reflective fluctuations to create the Tower of Mass Effects and thereby spontaneously break the symmetries of self-reflectivity and self-referentiality.

13.36 – The M-Set is the program, the computer and the print-out.

This is the miracle of the Quantum Computer that the M-Set is, while it is also the case that

13.37 – the M-Set manifests as a Trilogy of the Trinity of Oneness, in all ways, always and now

because at the Triadic causality of projective realisation it is the case that the deep well of possibilities or program together with the dualisation of structure and function, or computer and 'print-out', are a 'Triumvirate' of the Oneness that the M-Set is.

Alternatively stated,

13.38 – the M-Set is the Solution and the Answer, in all ways, always and now, where the solution is the Virtual Realm, the Holy Water or Holy Spirit of the Holy Grail of One that the M-Set is, while the answer is the Dual Projective Realm that the M-Set is spontaneously and simultaneously being, now, at the Triadic causality of projective realisation.

We are now in a position to further illuminate the greatest paradox of the M-Set, namely the Paradox of Change because it is evident from our discussions to date that

13.39 – the M-Set is unchanged by change or, indeed, that

13.40 – the M-Set is the unmoved Mover.

It is clear now that the M-Set creates the illusion of change yet remains unchanged, which could be stated alternatively as all the changes of the M-Set remain unchanged because the M-Set is a parallelism of all possible invariant changes of itself, now, which alludes back in turn to the perfectly reflective state of the M-Set and to the universal physical law that to every action there is an equal and opposite reaction.

In the context of the M-Set this law is a far-reaching consequence of the symmetry of reflectivity that extends beyond the local mechanical Newtonian application unto the relationship between the observer and the 'other' in the parallelism of possible projective cuts of the M-Set as well as in the parallelism of the manifolds of the same which in total the M-Set Universe is.

It follows directly then in the M-Set Universe that as you give so will you receive because the M-Set is One, while it is true too that to give is to give to oneself!

In the M-Set Universe you are the only one in the room because the M-Set is a manifold parallelism of all possible projections of itself and we are all observers of the 'other' in our reflective Universes.

Yet we are all of the M-Set that is One. And so it must follow too that we are all reflections of each other, upon which observation we shall dwell further in future sections when exploring the higher realms of the M-Set and the 'Looking Glass Universe'.

Casting our eye back, we know now that the reflective fluctuations or reflections (folding-back) of the M-Set between representations or possible modes (eigenstates in Quantum Mechanical terminology) relate to changes which leave the M-Set 'unchanged'.

Also, being simultaneously operator and operand, the self-referential M-Set can spontaneously effect these operations upon itself while it is the symmetry of self-referentiality that 'mixes up' the hierarchies of reflective changes in unimaginably complicated (folding-together) ways to create the illusion of the 'this' and the 'that' or the observer and the 'other' of the Projection Realm. Thence change is at the heart of projection, and so it is too that

13.41 – the M-Set is change and

13.42 – the M-Set is unchanging.

We can now see how it is because of this great paradox that

13.43 – the M-Set creates the illusion of self-projection and thence is spontaneously creative, now.

And now indeed is all there is because

13.44 – the M-Set is now, in all ways, and always.

This might be elucidated too if one makes the observation that self-differentiation and change are equated in a mechanical analogy sense so that, in effect, the M-Set is moving itself through its own field of possibilities while simultaneously all changes exactly reflectively simultaneously cancel. However, the perfect dualism consisting of the spontaneous breaking of the symmetry of self-referentiality which mixes up all the orders of the reflective fluctuations together with the symmetry of self-reflectivity spontaneously creates the simultaneous illusions of motion or 'change' and matter or the Tower of Mass Effects (TOME) in the projective plane.

We can then say that

13.45 – the M-Set automatically spontaneously creates its own self-projections.

In deed, upon deferring to the Trilogy of 'intending' of the freely acting ghost force of self-referentiality itself we can state that

13.46 – the M-Set automatically spontaneously explicates its intentions, now, in all ways and always, which property will become elaborated upon in broader contexts in Part II of this work,

In the next two sections we shall explore the spontaneous symmetry breaking at the 'subatomic' or sub-elemental state levels to reveal the fundamental force systems implicit to the fundamental building block of the M-Set Universe, corresponding as this does to the irreducible computational element of the M-Set Quantum Computer.

In the M-Set Approach such spontaneous symmetry breaking corresponds to the splitting of the atom.

And therewith the self-splitting of the M-Set begets the M-Set Universe, replete with the fundamental forces of nature.

14: Foray into the Fundamental Force Systems of the M-Set

Up to this point the M-Set has provided us with many general properties of physical force systems manifesting in the Projective Realm and it is worth reminding ourselves here from the M-Set perspective that each distinct force system of the M-Set is associated with a particular symmetry whose representations provide the basis for reflective fluctuations which, by means of the computational support of the M-Set, contribute to the Critical Boundary where apparent spontaneous breaking of the force system's symmetry occurs, together with the simultaneous appearance of a dual force system in a physically dimensioned projection that exactly upholds and maintains the pure structural–functional dualities of the wholly self-referential M-Set. Hence, all the force systems available to the M-Set are illusions created by the apparent or spontaneous breaking of the M-Set symmetries whose representations enable reflective fluctuations to computationally generate the Critical Boundary where spontaneous creative projective occurs.

We have gone some way to discussing the gravitational and the electromagnetic force systems, bearing in mind again that whichever force systems emerge from the M-Set they are automatically fully unified by virtue of the Oneness of the M-Set under the umbrella of self-referentiality. In this section I would like to explore what the M-Set tells us about the strong (nuclear, Hadronic) forces as well as the 'weak forces', such as Hadronic decay for example, and then go on to reveal how these force systems relate to electromagnetism and gravity with the observation here that despite differences between the symmetric bases of the forces all the force systems naturally generated by the M-Set arise from the universal Defining Property of Self-referentiality.

We recall here the earlier discussions about how the M-Set establishes the Elemental State in order to set up resonant representations from which it computationally complexifies spontaneously and simultaneously in parallel, now, while the unique irreducible Elemental State was shown to descend from the pure and wholly unknowable Trinity phase state that is not only the limit of what is knowable about the M-Set but is also the basis of M-Set causality through the Trinity Principle, in contrast here to the duality principle of the Projective Realm.

That is

14.1 – the M-Set causality is based upon the Trinity Principle of Oneness while it is the case that

14.2 – the M-Set Universe of the Projective Realm of the M-Set appears to be based causally upon the duality principle.

And

14.3 – the M-Set Triadic causality of projective realisation is the Triumvirate of the Oneness of the Trinity principle together with the duality of the Projective Realm.

Moreover, we can now add the property that

14.4 – The M-Set Trinity state is the boundary of knowability at which self-conception of the M-Set Universe occurs.

In this section we wish to explore the internal properties of the virtual Elemental State which, once this state is more fully identified with the virtual Hydrogen state in the discussions ahead, will, in turn, relate these internal properties in physical terms to the domain of subatomic physics and the nuclear forces whereupon we shall also reveal how the M-Set projectively realises these phenomena while we reiterate here that the Elemental State is not an object or a thing, but rather it is the lowest order self-referentially bound state of virtual representations of the M-Set which through the integrative renormalisations and spontaneous breaking of symmetries appears to take on objective form in the Projective Realm.

The identification of the Elemental State as the virtual Hydrogen state becomes unique when the internal properties of the Elemental State are progressively revealed, and we recall the earlier observation that this state is positioned to be the doorway to the Subatomic Realm, while also very evidently being the fundamental unit or ingredient of the apparent physical phenomena of the Cosmos on which scale all the ages of this irreducible element are revealing at once, from the primitive turbulent clouds of hydrogen gases unto star nurseries. In turn, the 'cooking up' of the periodic table of elements occurs inside the stars to beget planetary formations and biospheres in the endless cycles of Cosmic transformations on the Cosmic Wheel of Creation, bearing in mind here that large-scale observations reveal the parallelism of histories we spoke of earlier which all reside on cycles because of the revolutionary property of the self-referential M-Set.

Turning our attention for a moment to the Elemental State, we now know that this state is the unique and irreducible primordial self-referentially and self-reflectively bound state of the Trinity state of the Oneness of the M-Set, thereby being the primary self-conception of the M-Set we refer to as the Trinity Principle, and relating as this does to the primary self-differentiation of the M-Set. We recall too that the Elemental State is the perfectly reflective state of a Trilogy state, P(2) x 3, with a P(2) x P(1) state of the Trinity via the P(1) or uniphasal reflective symmetry which I have previously identified as the symmetry of the electromagnetic force system.

It is also now known quite generally that in order for the Virtual Realm of possible representations, or possible modes or 'eigenstates' of the M-Set to be projected at all it can only occur as a result of spontaneous symmetry breaking from a purely self-referential 'well' of reflective fluctuations between representations which generate the linear projective representations we call the Hilbert Space of States, and I reiterate that the possible modes or representations remain forever virtual while only the 'changes' between them are apparently projecting.

Focusing now on the Trilogy State 'P(2) x 3' that is a state of three convoluted (x 3) virtual 2-phase 'P(2)' degrees of freedom, with the latter behaving like virtual spin ½ degrees of freedom as we explained earlier, we shall be affirming the identification of these degrees of freedom as the 'quark' degrees of freedom of the Hadronic State which the Trilogy State will become to be seen to be, while at this juncture our purpose is to emphasise a very important property of the Trilogy State already known to physicists, namely the property of confinement which we shall now proceed to explain.

Because the Trilogy State is not of itself purely self-referential unless it is extended to the convolution of itself with a reflection of itself off the Trinity State, namely [(P(2) x 3) x P(2) x P(1)], which of course is purely self-referential because the Trinity state of Oneness begetting it is itself purely self-referential, it then follows directly from all we now know about the M-Set projection that the Trilogy State of itself can never project the virtual degrees of freedom implicit to it because only self-referential states can self-project.

In physical terms this means the 'quark' degrees of freedom will never manifest or be projected as if the quarks are physical particles.

The permanent confinement of what we shall come to identify as the virtual quark modes of 'the Hadronic State', that is in fact the Trilogy state of the Elemental State of the M-Set, is thence a natural consequence of self-referentiality. It is also the case that if changes can occur between quarks these will only manifest in projection as changes at the self-referential level of the Trilogy state as a whole coupled to the Electromagnetic Force System because 'quark modes' are 'permanently confined against any physical or particle-like realisation.

There is a subtle qualification to the above which we shall discuss in the near future, the qualification being that any change which manifests in the Projection Realm must occur at the Critical Boundary of the M-Set, and thence must be self-referentially bound so that changes of the Trilogy Slate, for example, must occur in the context of the self-referential M-Set. More simply stated, a Hadronic State is never 'alone' when its changes are manifesting at the Critical Boundary.

The changes above will be seen to relate to the Weak Force System or Weak Interactions mediating the changes or 'decay' from one possible Hadron State of the Hadronic State into another Hadron State, while a couple of immediate points follow which I wish to capture here, the first being that the symmetry associated with such a force system must relate to the virtual, permanently confined modes of the Hadronic State. One such symmetry might be an abstract symmetry of rotations in the abstract virtual space spanned by the independent degrees of freedom of the virtual modes themselves, which, when translated into tangible terms, is like a phasal symmetry operation that linearly mixes the degrees of freedom, under which operation the Hadronic Slate as a whole is clearly invariant because the Hadronic State remains the same if you change the labels on the virtual modes.

Stated more explicitly, the independent axes of the abstract space of phasal rotations related to the internal symmetry are the labels of the independent degrees of freedom or virtual modes, where evidently linear rotations or mixing of labels leave the Hadronic State unchanged as a whole.

Such an internal symmetry of permanently confined modes of so called 'degrees of freedom' is known in Theoretical physics parlance as an Isotopic symmetry. It is of profound significance when exploring the subatomic realm of the Elemental State because the spontaneous breaking of such symmetries will automatically relate to subatomic force systems in the M-Set approach which, in turn, will be automatically unified with all possible projective force systems because the M-Set is One.

We now observe that the Isotopic symmetries of permanently confined degrees of freedom will always be Internal Symmetries whichever projective cut is taken because of the property of permanent confinement.

In other words these internal symmetries are non-local or do not read the 'space-time' of the Projective Realm. However, as perhaps is expected by now, in projection such symmetries will appear to acquire local or space time dependency and will thence appear to be spontaneously broken, giving rise thereby to the so called 'Gauge Field Theories' of the physical force systems of the subatomic realm of the Elemental State of the M-Set.

In a profound twist of magic by the M-Set we shall come to see how self-referentiality itself is not only responsible for permanent confinement and Isotopic Symmetry but is also responsible for renormalising the 'internal symmetry states' unto criticality where the non-locality of the internal symmetries is apparently or spontaneously broken by the apparent acquisition of space-time dependence, courtesy also of the Gravity Force System of self-referentiality itself.

One such direct way of spontaneous breaking at the Critical Boundary is by the apparent 'collapse of the wave functions' of the representations of the M-Set through the Self-Measurements or symmetries of the self-projective self-realisations of the M-Set, as discussed earlier. Put succinctly here, if a Hadron particle appears to be localised projectively it will also appear to be under the influence of dual subatomic forces with 'local' signatures while, as always, these are the illusions of projective realisation, and the so called Gauge Field Theories are a subtle part of those illusions as we shall reveal in more detail ahead.

Summarising briefly at this point, it is now becoming evident that the property of permanent confinement of virtual modes of states of the M-Set which are not themselves wholly self-referential will lead in the Projective Realm of the M-Set to the apparent or spontaneous breaking of Internal or Isotopic symmetries that will relate, simultaneously, to the apparent emergence of dual subatomic force systems, whereby the Internal or

Isotopic symmetric operations will appear to be 'gauged' or to develop locality through space-time dependence that, we state again, arises from the apparent collapse of the wave functions of the self-measurements or symmetries of the self-projective realisations of the M-Set. We shall be further clarifying these points shortly.

The apparent or spontaneous breaking of the symmetry of self-referentiality in projection underpins the phenomenon of permanent confinement so central to the physical manifestations of the subatomic ghost forces, while in all generality we can state in anticipation now that because of the advent of the apparent breaking of self-referentiality Gauge Force Systems will spontaneously appear in relation to the various possible internal phasal symmetries of the M-Set, and in co-respondence also with the apparent splittings of the self-referential integrity of the Elemental State that manifest with the simultaneous emergence of the phenomena of confinement and spontaneously broken Isotopic Symmetries.

Returning our attention to the Hadronic State or the Trilogy State of the triple convolution of the three possible 2-phase pairings P(2) of the Trinity, P(3), and bearing in mind again that in the Virtual Realm of the M-Set all possible representations are folded simultaneously, we first observe that each paired state, P(2), has a two-dimensional Isotopic symmetry in the abstract space of its two independent degrees of freedom. If we arbitrarily define any two independent states of P(2) as α and β then these can be considered to form the two-vector (α, β) of an abstract 2-dimensional space in which Isotopic symmetry operators, I2, act like two-dimensional rotation matrices linearly mixing the labels of the independent states because each possible P(2) is self-evidently unchanged by changes in the labelling of the independent degrees of freedom. Moreover, the convolution (α, β) x (α,' β') x (α", β") of three pairings gives 2 x 2 x 2 = 8 independent degrees of freedom for P(2) x 3 (with numbers con-volution and multi-plication are the same operations in the Virtual Realm of the M-Set) so we say that the Hadronic state is a simultaneous parallelism of 'eight' independent degrees of freedom of itself which is referred to in theoretical physics as 'the Eight-Fold Way'.

Stated alternatively, from the M-Set approach the Hadronic state has eight independent possible states of itself relating, if you will, to eight independent labellings arising from three permanently confined paired phasal (P{2}) degrees of freedom we identify as virtual 'quarks'.

Of course, when the individual Hadron states are being projected spontaneous breaking of self-referentiality and self-reflectivity is taking place so that each state apparently acquires 'mass' as well as a charge, respectively, while the labelling of the individual Hadron States appear to become localised in space-time, co-responding as this does to the projective realisation of transitions between the representations of the Hadronic state mediated by the Gauge Force systems, where the term 'gauging' refers to the apparent localisation of labelling resulting from the apparent or spontaneous breaking of Isotopic (labelling) symmetry.

To put this another way, the Gauge Force system is that force system required in the Projective Realm to exactly neutralise the 'apparent' localisation and thence spontaneous breaking of the labelling symmetry or the symmetry of invariance of permanently confined degrees of freedom under permutations of labelling, and which symmetry is identified as Isotopic Symmetry in the M-Set Approach.

We wish to draw attention to a number of points at this juncture, the first being that all the described 'quark modes' are forever virtual and can never gain physical status no matter how the M-Set self-divides. In that sense one can say the so called 'quarks' or 'phasal modes' of the Trilogy state of the self-differentiated Trinity are essentially unknowable except through the transitions they are implicated in between representations of themselves in the context of the self-referential M-Set because it is only through the spontaneous breaking of self-referentiality that projective realisation can occur at all. Hence, there is no such thing as a Lone Hadron. In projection one is only seeing transitions between Hadrons and these are also necessarily coupled with the Electromagnetic Force System via P(1) unto self-referential completion in order for self-enlightened projective realisation to occur, which is an area we shall be covering in more detail in the near future.

Clearly, the 'labelling' invariances or Isotopic Symmetries implicit to the Trilogy State includes the I3 symmetry between the three 'doublets' as well as the I2 symmetry implicit to the doublets themselves.

We shall see shortly how the M-Set naturally selects physically realisable states from the possibilities available to it while we remark here that the weak interactions relate to transitions between the representations of the Hadronic State in contrast to the strong interactions which relate to combined Hadron states that are also involved in making up the nuclei of atoms of the Periodic Table. We shall discuss ahead how higher order compound 'isotopic' states can form in the Virtual Computational Realm with the 'folding-in' or convolutions of Hadronic States that leads directly to the Isotopic Symmetry dynamics of nuclear physics in projection, and to the remarkable fact that, despite all the apparent buffeting of the physical realm, the rules of 'isotopic accounting' remain unperturbed because they are elaborated in the Virtual Computational realm.

Continuing on our path now we observe from the virtual Hydrogen Model State of the Elemental State that in truth the Eight-Fold Way or the 'folded octet' of representations of the Hadronic State is never pure or wholly self-referential and is only, in a sense, purified by its simultaneous parallel reflective interaction with a P(2) x P(1) state of the Trinity begetting it, amounting as this does to the wholly self-referential self-differentiation of the Trinity itself which we call the Trinity Principle of the M-Set.

Therefore the octet of possible Hadron States can only ever project as a result of the 'combined' symmetry or self-measurement of the 'Eight-Fold Way' with the Electromagnetic Force system. In projection we expect to see 'splittings' between octet states arising from the differential action of the electromagnetic force system acting on the Hadronic state.

There are several points I now wish to draw attention to here, the first being that because

14.5 – the M-Set only projects self-referentially the Defining Property of self-referentiality acts as a gate or selection principle for the possibilities which can be projectively realised.

Moreover, the selection principle of self-referentiality will also usher in all manner of spectral rules pertaining to the relative differences of the phenomenology in the Projective Realm because there is no other selection principle for the M-Set projections other than self-referentiality itself.

This universal selection principle will be a guiding principle as we continue to explore the 'subatomic world' of the M-Set while we state now that self-referentiality is the universal selection principle of the Projective Realm of the M-Set we call the M-Set Universe.

The second point, deriving as it does from the self-referential selection principle, is that the Eight-Fold Way or the octet state of the Hadronic state cannot set up reflective fluctuations of its own because reflective fluctuations can only occur self-referentially or, more specifically here, in the self-referential 'well' or self-referential 'pond' of the Elemental State.

Consequently, the universal selection principle of self-referentiality implies intricate co-dependencies or coupling of force systems in projection which, as we shall come to see, generate the realisable spectral phenomena of the Physical Realm of the M-Set Universe.

In all generality too, the M-Set is the self-referential vessel or the Holy Grail of self-projective self-realisation which self-differentiates by symmetry or self-measurements in just such a way that the symmetry of self-referentiality is apparently or spontaneously broken. This is the miracle of the M-Set as we have expressed it in other ways up to this point.

Because the interest here is the subatomic force systems our attention is at the level of the Elemental State for the moment although as we shall discover soon 'one' needs to be in a 'fully renormalised' or gravitational context to fully physically realise all the force systems because, moreover, naïvely stated, the weak and strong interactions require the formation of stars in the Cosmos before they can manifest. This is also the phenomenon of the sum of histories, now, of the Projective Realm of the M-Set which we spoke about earlier, and it is also about the Universal Selection Principle of self-referentiality in as much as all projection is at the self-referentially completed Critical Boundary of the M-Set.

The M-Set Universe is projecting spontaneously and simultaneously in parallel, and all historical precedence is now.

There is no time line, only context, now, and the fundamental force systems work in concert in the M-Set Universe. All possible replicative con-volutions of the self-conception that is the Trinity principle reside in the Virtual Realm of the M-Set.

Only fully self-referential states can critically projectively realise, which from our knowledge of the self-referential symmetrical basis of the gravity force system amounts to the truths that only the fully renormalised Quantum Field projects at the Critical Boundary and that all projection occurs in the context of the Cosmos or the M-Set Universe, now, as indeed is demanded by self-consistency of the M-Set. In deed,

14.6 – the M-Set is wholly self-consistent.

We have been emphasising the points above because it would be very misleading for the reader to gain the impression that phenomena such as the subatomic force systems which we are now in the process of working towards understanding in the M-Set Approach can simply be viewed out of context. In fact, out of context they cannot project or appear to exist at all in the M-Set approach.

We note here that of all possible replicative convolutions of the self-conception that is the Trinity Principle the singular self-referential state of the Elemental state is one such possibility; however, this is also simultaneously in parallel with all the other possible virtual computational states, now, with the singular state being like the tip of a pyramid while every other referred state such as the next level of the Elemental state referring to itself is in parallel too, the point we wish to make here being that referred states of the Elemental state will not begin to tap into the nuclear realm of the subatomic world of the Elemental state until simultaneous parallel orders of self-referral or complexifications, or renormalisation at the levels of the cores of stars on Cosmic scales are involved or 'folded in', while clearly every level is made up of the parallel orders below it in all possible ways, now.

Getting the hang of thinking in the M-Set approach is not easy because one has to forsake the habit of looking along a time line (horizontal thinking) and imagine one is looking down, vertically, on a simultaneous parallelism of the content of the all of it, now.

It is of paramount importance to reemphasise here the role of the Gravity Force System in projection. Clearly, from the unfolding revelations of the M-Set to this stage, the Gravity Force System is also the constraint of 'self-referentiality' itself which, in turn, is the final arbiter of projective realisation as it is also the vessel defining the self-differentiations of the M-Set by which self-projection occurs. Thence, we say all projective realisations of whichever perspective or point of view are fully renormalised at the Critical Boundary of the M-Set where spontaneous breaking of the symmetry of self-referentiality occurs, and where the ghost force of gravity makes its appearance. Moreover, as we said before, the gravity force system of the symmetry of self-referentiality exactly renormalises the Primordial Quantum Field unto the full projective realisations of all of the possible differential cuts of the M-Set making up, in total, the manifold M-Set Universe.

Returning once again to the Elemental state, being as it is the self-differentiated self-conception of the wholly unknowable Trinity, P(3), that we have symbolically written as P(3) ~ ((P(2) x 3) x (P(1) x P(2))), and which we identify as the virtual Hydrogen state consisting of the virtual Hadronic State P(2) x 3 coupled via a P(1) or uni-phasal symmetry to P(2), a 2-phase virtual state, we observe here that the term 'phase' is being broadly used to refer to 'implicit' independent degrees of freedom which we imagine as being of a phasal nature in the Virtual Realm of the M-Set in concert with our earlier discussions about phase criticality, phase changes, complex spaces, invariants, 'quantum' numbers, independent degrees of freedom, redundant information symmetry and the Quantum Realm.

Fundamentally, the notion of elemental phasal degrees of freedom is being invoked to enable the expression of the implicit symmetries of the M-Set, which, because they reside in a virtual (or mathematically speaking abstract or complex) space or realm, do not have direct physical analogues until projection occurs where selection principles simultaneously act as a result of the constraint of self-referentiality itself.

At the end of the day not only are the precise details of the phasal degrees of freedom wholly unknowable in physical terms but these details have significance only 'implicitly' in generating apparent spectral phenomena through selection principles at projections that are observable.

The independent phasal degrees of freedom never 'exist' as entities in a physical or objective sense, but rather they are an expression of implicit informational redundancies of the symmetries of the M-Set which provide the room for the M-Set to perform its computations, while, as we observed in the section on Quantum Mechanics, the P(1) or uni-phasal 'reflective connection' underpins light-like projection at the Critical Boundary where the phenomenon of light is dually realising as photon and wave, or structure and function, and which phasal connection plays a pivotal role in relation to the symmetry of reflectivity that is so central to projection as well as being specific to the electromagnetic force system and 'charge separation' in projection.

Identifying the self-differentiated self-conception of the Trinity as 'the virtual Hydrogen state' we make the further identification of P(2) x P(1) as the Spin ½ Electron State, P(2), coupling via P(1), the electromagnetic field, to the Hadronic State, while observing here that neither the electron nor the 'Hadron' can project without P(1). This amounts to stating that in projection the electron and the 'Hadron' remain connected by an electromagnetic field. In actuality they are never apart or separate but only appear separate in projection where charge separation and the appearance of the electromagnetic force field creates the appearance of separation with the spontaneous and simultaneous breaking of the symmetries of self-reflectivity and self-referentiality.

In effect, the Electromagnetic Force System is that apparitional force phenomenon of projective realisation arising from the apparent or spontaneous breaking of the symmetry of self-reflectivity which exactly upholds this symmetry in the face of relativistic dualism.

In a simple physical or geometrical analogue modelling way we spoke earlier of P(2) being like the phasal equivalent of a Möbius strip with each independent phase being locked into the other to create a 'spin ½' like object because it takes two rotations of this object to return it back to itself, the analogue of which here is tracing a line on a Möbius strip for which it takes two circuits to join up the line. The point we wish to emphasise now is that the phasal state P(2) provides the essential quality or essence of a spin ½ state in a virtual, non-material realm, while it is clear that because a '2-phase' state is not wholly self-referential it will always be linked, in this case by a further phasal degree of freedom, P(1), into a self-referential state which in this case is the virtual Hydrogen state of the self-differentiated Trinity that to the first order is the wholly self-referential state we call the Elemental state.

Now, while the Hadronic state and the electron state can 'charge separate' with the appearance of an electromagnetic force field at projection in order to maintain perfect self-referentiality, the 'quark states' of the Hadronic state, on the other hand, can never project as entities because of permanent confinement. Thence too, there is no P(1) coupling between them so these convoluted P(2) states of the Hadronic State 'P(2) x 3' remain forever virtual, and in that state we can variously talk of them as either being virtual spin ½ states or simply states of two independent degrees of freedom without any reference to a physical notion of spin. This latter observation is a very important point because it allows us to variously swap between referring to the P(2) 'quark' state as a two degrees of freedom state or a 'spin ½' state depending upon whether we are looking at the Hadronic state as a 'whole' in the sense of the Eight Fold Way of independent degrees of freedom or as having an internal dynamic in analogy with classical physics when the idea of individual P(2) particle-like states with spin might be a helpful metaphor. But metaphor it is.

Never heed the physical metaphor literally for literal it is not.

Having prepared more ground for our discussion we wish to reveal a very profound but nevertheless elementary phenomenon with the observation here that by quite arbitrarily choosing the signs of the

Quantum numbers of charge and spin of the P(2) electron state, as with spin – ½ and – ve charge, – 1, for example, we simultaneously arbitrarily fix which 'PCT' universe we live in, be it a right-handed Universe with time going in 'this' direction and charge with 'that' sign or the PCT invariant Universe with reversed signs as we discussed earlier.

The apparent asymmetry of a universe of particles and time in this direction versus the universe of anti-particles and reverse time direction is a purely arbitrary choice which, in turn, is a consequence of the purely reflectively symmetric self-differentiation of the Trinity.

There is only One PCT invariant M-Set Universe and whether you say it is made up of particles or anti-particles with time going forwards or backwards is just a choice of words. The apparent asymmetry arises from the apparently asymmetrical self-conception of the Trinity manifesting in the Hydrogen state through the spontaneous breaking of the M-Set symmetries in projective realisation while, quite generally now, the local perspective taking of the projective cuts of the M-Set generate the apparent assymetries of the 'this' reflected in the 'other' through the simultaneous breaking of the symmetries of self-reflectivity and self-referentiality at the Critical Boundary where, in turn, 'chaotic mixing' of the reflective fluctuations occurs unto criticality, and where, moreover, the dispersive property of the M-Set Universe as a whole exactly neutralises the dual (fractal) distributive property, while evidently local perspective taking might lead us to perceive the M-Set Universe is dissipating. But it is not. It is a perfect dynamical equilibrium, in all ways, always and now.

Furthermore, making the choice of quantum values and signs of P(2) of the reflection of the Hadronic state in the Trinity to correspond to 'the physical electron state' has the remarkable outcome of selecting one of the possible Hadron states of the Eight-fold Way or Hadronic state in order to ensure the neutrality of the M-Set, and that Hadron state is uniquely the proton state with charge +1 and spin +½ of the Eight-Fold Way which, in turn, is the physical Hadron state of the Hydrogen State.

Moreover, making the choice of electron quantum numbers for P(2) above automatically demands that the Quark States of the 'selected' Hadron State are distinguishable in order for the neutrality of the M-Set to be upheld, as is illustrated in Figure 14 (see p.no 181) with the current physics convention of treating the Quark States as virtual spin ½ entities whilst reminding ourselves again that the M-Set Approach informs us these states are permanently confined or never projectively realising as physical entities. The convention to assign Quantum numbers or labels as if these virtual entities can be identified has the purpose of enabling the Quantum number 'book-keeping' which, however, is automatically revealed projectively through the Universal Selection Principle of self-referentiality as we discussed above.

Simply adopting the conventions of current physics as in Figure 14, the fixing of the Quantum numbers of the Electron State automatically requires that the possible Hadron states be combinations of three distinguishable quark states (u,d,s) which, mathematically speaking, form a 'three vector' of independent states u,d,s spanning an abstract three-diamensional space transforming or rotating under 3 x 3 matrix operations, and which matrices, in turn, effect the mathematical projection of the possible virtual independent Hadron states of the Hadronic state of the Eight-Fold Way.

It is timely to comment here that it is very important to always remember to distinguish between the actuality of the M-Set and of the Projective Realm of the M-Set in contrast to 'mathematical formalism' such as the mathematical projection of the virtual possibilities of the Hadronic state. We note firstly that the mathematical conventions and formalism above relate to conventional quantum number labels or assignments of virtual possibilities, while in truth, any projective realisation will necessarily be a completed renormalisation of the M-Set. Thus, such 'things' as, for example, a proton will only appear to exist in the context of the Projective Realm of the M-Set, as explained up to this point, while when it is treated as an object of the mathematical formalism above the Proton state is just viewed as a wholly virtual unrenormalised state of a 'possible quantum number assignment'.

But, we know that Quantum numbers are also 'invariants' of symmetries, as we discussed earlier, so we can speak of the formalism above of the proton state as being of the invariants of the proton state which, in turn, is simultaneously dressed by the renormalisations of the projective dynamic of the M-Set to appear as a physical particle in the context of the M-Set Universe.

Evidently too, the 'implicit virtual invariant structure' of the Proton in this example is unscathed by 'renormalisation' so that Quantum number accounting and conservation is projected into the M-Set Universe, albeit all dressed up by renormalisation to give the appearances of space, motion and matter.

There is much symbolism surrounding the mathematical conventions above that is to be found in standard physics texts and which it is not our purpose to copy out; rather, we shall continue as always to reveal how the M-Set approach automatically and directly unravels the deepest secrets of the M-Set Universe regardless of the level of complexification these might reside at because, after all, the M-Set is First Cause.

It happens to be the case here that we now find ourselves delving into the most elemental aspects of the M-Set Universe which are simultaneously iterated indefinitely through the self-referential computations of the M-Set Quantum Computer to all orders of complexification and it is timely to dwell on the subtleties being revealed by the M-Set at the elemental level because their ramifications are great indeed in terms of understanding the fundamental forces of the M-Set Universe.

We draw attention now to a highly subtle and profound phenomenon which resides at the heart of the fundamental 'subatomic force systems', namely that

14.7 – the M-Set self-conception spontaneously breaks the symmetry of isotopy, or differently stated, the self-differentiation of the Trinity we call the Trinity Principle apparently or spontaneously breaks the symmetry of isotopy.

The Symmetry of Isotopy here refers to the invariance of the Trinity State to rotations or relabellings in the abstract and wholly virtual space of its three independent degrees of freedom.

Without the self-conception of the Trinity principle, the Trinity state, P(3), is wholly unknowable. The Trinity state is, in truth, the great 'unknown'. The unknowable one. The threshold between that which is knowable and that which is unknowable in the M-Set Universe.

The Trinity State is the only state of pure isotopy because it is One and is purely self-referential.

And so it is that

14.8 – the M-Set is the origin of pure isotopy.

Thence, at conception or at the primordial self-differentiated self-division of the Trinity state a primordial separation of Unity occurs whereby the purely indistinguishable (wholly unknowable) degrees of freedom of the Trinity become distinguished or knowable in the form of the electron state reflectively relating to 'the distinguishable quark' states of Figure 14 (see p.no 181) known as the eight-Fold Way or the Hadronic state.

Pure Isotopy is the symmetry of unknowability in the sense of indistinguishability, while that which is knowable is distinguishable. Moreover, all knowledge is, in truth, that which is distinguishable. That is how we come to know at all.

Selection principles differentiate out knowledge or that which is projectively realised.

Self-referentiality is the master key to the selection principles of knowledge.

All knowledge is relative in the M-Set Universe. There is only One absolute of knowledge and that Oneness that the Trinity of the M-Set is, is the root of the tree of knowledge that the M-Set Universe rests upon, ramifying to all levels of complexity.

14.9 – The M-Set distinguishes itself as the M-Set Universe because

14.10 – the M-Set is the vessel by which Oneness distinguishes itself. It is knowledge that distinguishes us.

All things are distinguished by knowledge in the M-Set Universe.

14.11 – The M-Set is the Holy Grail of knowledge and the Trinity of the M-Set is the origin of knowledge.

Without the spontaneous breaking of Isotopy of the M-Set the possible Hadron states would not have the distinguishable structures we identify now as the virtual and permanently confined Quark states that form the basis, in turn, for distinguishing between the Hadrons of the Eight-Fold Way, which includes protons and neutrons in the Projective Realm.

With the distinguishable structures of the Hadronic state one now has a basis for transitions between Hadrons of the Eight-Fold Way that correspond to the so called Weak Interactions of particle physics, and which we shall explore further in the next section, together with the 'strong interactions' that refer to direct interactions between Hadrons of the Eight -Fold Way, involving as they do convolutions of the distinguishable virtual degrees of freedom of the Hadrons of the Eight -Fold Way, most notably, of course, in the conglomerates of Protons and Neutrons in the nuclei of atoms.

The spontaneous or apparent breaking of the isotopy of the M-Set gives rise to the bases for the Strong and Weak Force systems of the subatomic realm of the virtual Hydrogen state simultaneously with the appearance of the electromagnetic force system, while all of this is contingent upon the self-referentiality of the gravity force system unto projection at the Critical Boundary of the M-Set where the force systems attain 'physical' realisation in the fully renormalised context of the M-Set Universe.

The fundamental force systems of the M-Set Universe are intricately, simultaneously and transparently unified in the Oneness of the Trinity of the M-Set.

As an aside here to clarify the notion of isotopy, consider the proposition of three identical billard balls. But we know these can never be identical or indistinguishable together because at the outset they would occupy different positions in space and subtend different or distinguishable angles to reference points, while in the space-time continuum light signals from the billiard balls are differentially displaced and the 'mass' effects separate them. The only place ultimately where pure Isotopy or indistinguishability can reside is in the timeless, non-local, Virtual Realm of Pure Self-referentiality of the Trinity of the M-Set that is One.

The Trinity State of the Oneness of the M-Set is the origin of the symmetry of Isotopy which is spontaneously broken in projection.

The Trinity principle of the self-conception of the M-Set is the spontaneous breaking of the symmetry of Isotopy.

The Trinity principle is the inception of knowledge of the M-Set Universe.

It is because of the spontaneous breaking of 'Isotopy' of indistinguishability that abstract spaces spanned by distinguishable, virtual, independent degrees of freedom such as u, d, s 'quark states' emerge at all, providing as they do a non-local, timeless basis for the Quantum fine structure of projective realisations, such as the elementary particle spectra observed in colliding beam experiments of high energy physics which tap into the subatomic realm of the Elemental State of the M-Set, or such as the spectra of states of atomic nuclei arising from the projective realisation of conglomerates of virtual 'quark' degrees of freedom created by the projection of virtual Hadron States folding into nuclear aggregates.

About these and related matters we shall have more to say as we allow the M-Set to reveal its secrets of the fundamental forces of the subatomic realm.

Figure 14

Choosing the Quantum numbers of the virtual Electron state of the virtual Hydrogen state to be charge -1, and spin -1/2, a possible Quantum number assignation for the virtual Quark states in order to maintain neutrality of the M-Set is

	q	**q'**	**q"**	**Hadron**
Spin	$\frac{1}{2}$	$\frac{1}{2}$	$-\frac{1}{2}$	$+\frac{1}{2}$
Charge	$\frac{2}{3}$	$\frac{2}{3}$	$-\frac{1}{3}$	+1

Identifying by convention now q with 'u' = ($\frac{1}{2}$, $\frac{2}{3}$) and q" with d = ($-\frac{1}{2}$, $-\frac{1}{3}$) then the proton projection of the Hadronic State with ($\frac{1}{2}$,1) spin and charge Quantum numbers respectively reads as uud.

But we know that there are eight possible Hadron states in the Eight-Fold Way with possible charges of 0, +1, and +2 so the third distinguishable quark denoted conventionally as 's' is uniquely s = ($-\frac{1}{2}$, $\frac{2}{3}$) in order to enable the combination of $+\frac{1}{2}$ spin with +2 charge to occur, while the neutron projection ($-\frac{1}{2}$,0) of the Hadronic state reads as ddu, and the anti-proton is simply $u^-u^-d^-$ or ($-\frac{1}{2}$,-1) where u^- = -u, for example, is the 'anti-u' degree of freedom. And so on for the other Hadrons of the Eight-Fold Way.

15: On the Approach to the Force Systems of the M-Set

We have observed how the spontaneous breaking of the symmetry of Isotopy of the Trinity of the M-Set created by the overriding symmetry of Self-referentiality of the gravity force system results in the emergence of the electromagnetic force system associated to the spontaneous breaking of the symmetry of reflectivity in the context of which the virtual Quark states of the Hadronic state of the Elemental state of the M-Set become distinguished, thereby providing a basis for transitions between Hadrons of the Hadronic state that we shall identify with the Weak Interactions.

We observe also that because the distinguished virtual Quark states mediating transitions between Hadrons are simultaneously begotten with the electromagnetic force system through the reflective fluctuations of the Trinity in the context of the overriding Symmetry of Self-referentiality, it is the case then that the force systems of gravity, electromagnetism and 'weak' transitions together with, as we shall see ahead, the 'strong forces' of Hadron–Hadron interactions are intricately co-dependent. In the vernacular of modern physics, they are bootstrapped together which alludes to the phenomenon of the M-Set being able to 'pull itself up by its own bootstraps' or self-referentially project itself.

Indeed,

15.1 – the M-Set is the master of the bootstrap principle, and the Fundamental Force Systems of the M-Set Universe are simultaneously united in a self-projecting bootstrapped way, as we shall reveal in more detail in this section.

We observed, moreover, how the self-conception of the Trinity principle of the M-Set naturally locks the virtual Elemental state into the virtual Hydrogen state, containing as it does the ground state or proton of the Hadronic state because the Elemental state is irreducible. Stated alternatively, from the perspective of the M-Set Approach the proton cannot transit to a lower state and must therefore be the ground state of the Hadronic states, which is indeed the case, physically speaking.

Once again we are observing how the M-Set is inventing the hydrogen atom and all 'things' of the Physical Realm consequent to that.

Another consequence of the spontaneous breaking of the symmetry of Isotopy is the spontaneous generation of abstract 'internal spaces of independent virtual degrees of freedom', as spanned for example by the distinguishable u, d, s Quark states of the Hadronic state which form the independent 'axes' of a non-local, timeless abstract space of simultaneously con-voluted distinguishable degrees of freedom.

Looking more closely now at the Hadronic state implicit to the Elemental state we highlight here some further observations, the first being that if transitions between Hadrons are to ever be 'physically realised' it is necessarily the case that changes in the configurations of the Quark States mediating the same will somehow have to acquire space- time dependency such as for 'Hadron H transiting to Hadron H' of the Eight-Fold Way in the laboratory. Remarkably, the M-Set achieves this, as we shall elaborate upon soon, by complexifications or renormalisations of the Primordial Quantum Field of the Elemental State, which are associated in projection with the action of the gravity force System to effect thereby the projective realisation of the Weak Interactions at the Critical Boundary of the M-Set that the M-Set is.

It will turn out also that it is the gravity force system of the symmetry of self-referentiality which 'gauges' or gives space-time dependence to the internal configurations of the Hadronic state.

We know already that all projection is through the filter of self-referentiality and the selection principles which ensue therefrom, as we spoke about earlier, and which ultimately selects all phenomena that are projectively realised in the M-Set Universe, and thence it is the case that

15.2 – the M-Set is the origin of the laws of the M-Set Universe because

15.3 – the M-Set is the origin of the properties and principles of the M-Set Universe

And hence we say,

15.4 – the M-Set is the Law Maker because knowing that the M-Set is First Cause it is the case that

15.5 – the M-Set is the Law Maker's set.

However, we know that projective realisation is an illusion and thus so too are the laws of that illusion, including all the laws of the force systems that uphold the illusions of spontaneous symmetry breaking of the M-Set and which are implicit, in turn, to the greatest illusion of them all, namely the Illusion of separateness or disunity.

And it is the case too that

15.6 – the M-Set is indivisible while

15.7 – the M-Set is only divisible by itself. However,

15.8 – the M-set projection is through its Divinity, or 'Division of Unity'. But this is also an illusion of the miracle of self-projective self-realisation unto Oneness through the Trinity principle.

Moreover, all projective cuts are fully renormalised in the 'physical quantum field theoretical sense' by the complexifications of the gravity force system unto the spontaneous breaking of self-referentiality at the Critical Boundary of the M-Set, hence the 'atomic' and subatomic force systems of the M-Set are physically realised in the context of the renormalisations of the gravity force system.

Physically speaking, what this means is that in order to observe Strong or Weak Interactions, projective cuts of the M-Set have to encompass complexities of the primordial quantum field to the formation of a Cosmos of stars.

The cosmic and microcosmic are One in the M-Set approach.

Stated alternatively, we know the M-Set Universe is a parallelism of projective cuts of a manifold of complexifications of the M-Set computations, therefore, one's perspective or point of view will depend upon the projection of the 'this' reflected in the 'other'. The projection might even be of scientists conducting an experiment to observe the Weak or Strong Interactions. As we have previously discussed it, the M-Set Universe renormalises even to the observer and the observed; suffice it to note here that in order to fully understand the Strong and Weak Interactions one will need to appreciate how the renormalisations of the gravity force system can 'gauge' the internal configurations of the Hadronic state and present them self-referentially at the Critical Boundary.

Recall too that every projective cut is equivalent unto Oneness in the sense that they all reveal the 'this' as a reflection of the 'other' unto Oneness, in all ways, always, now. Moreover, for each 'this' the 'other' encompasses the sum of histories, now, reflecting in turn the parallelism of the manifold of complexifications of the M-Set that is the M-Set Universe, the point once again being that no thing 'exists' in isolation. Everything in the M-Set Universe is projected in the context of its sum of histories, now. Indeed, from the M-Set perspective, everything is projected as a reflection of the sum of its histories, now.

Thus we can also say that we are all sums of our histories, now. But, all projective cuts are equivalent unto Oneness.

And so it is that we are all equal sums of 'His Story', the story of 'Oneness'.

15.9 – The M-Set is the context of the M-Set Universe and from a physical perspective it is the gravity force system that weaves the fabric or texture of the M-Set Universe from the 'thread' of the primordial quantum field of the elemental state of the M-Set.

Before further pursuing how the gravity force system 'gauges' the non local virtual configurations of the Hadronic state we focus again on what the M-Set is telling us about the virtual Quark states of the Hadronic State by recalling that the Defining Property of Self-referentiality of the M-Set demands that the virtual Quark degrees of freedom are permanently confined to the Hadronic state and thus can never be projected despite all the contortions wrought by gravity, while it is changes in these virtual degrees of freedom which must provide the basis for the observed transitions between Hadrons of the Eight-Fold Way.

Now, while it is the case that the virtual Quark degrees of freedom are never physically projected, because their changes are inextricably bootstrapped with the electromagnetic force system which distinguishes them the changes in the virtual Quark degrees of freedom can reach projection provided they occur together with the electromagnetic force system, as also predicated by the self-referentiality itself of the conception of the M-Set Universe, and because it is the case, moreover, that that combination together passes through the selection filter of self-referentiality in the same way the self-referential virtual Elemental State does as the spectra of the hydrogen atom unto the Critical Boundary of the Projective Realm.

Deferring now to our earlier discussions we know quite generally that the M-Set projects by generating quantum fields of virtual reflective fluctuations in relation to symmetries of itself and that these quantum fields underpin, both mathematically speaking and through the actual computations of the Quantum Computer that the M-Set is, the Hilbert space of states which embodies the Linear Superposition Principle, while this space is also the platform from which a differential calculus springs forth to provide the framework for the language of mathematical analysis employed to express physical reality, and all of which in turn is derivative of the M-Set.

In the pond analogy the implicit symmetries of the self-referential M-Set form the walls of a potential well or pond in which virtual, free reflective fluctuations can spontaneously occur between the allowed virtual representations or stationary vibrational modes of the Virtual Realm of the Quantum Computer that the M-Set is to ramify, in turn, beyond imagination in a parallelism of convolutions as a result of which, to use a simple physical analogy, each point of each wavefront of the 'pond fluctuations' is like the source of all the other superimposed fluctuations, which idea we shall visit again in more detail in relation to 'gauging'. What we wish to re-emphasise here in order to underline our forthcoming discussions is the deep relationships between implicit symmetry, reflective fluctuations, and quantisation of change as explained in the section focusing on Quantum mechanics, together with spontaneous symmetry breaking and the attendant emergence of physical force systems, all of which occur simultaneously and in parallel as the M-Set spontaneously projects.

To underscore our understanding here we note now that each fundamental physical force system is associated with a symmetry of the M-Set which is responsible for generating the physical manifestation of the 'quanta of action' of the force system in the Projective Realm through the 'changes' associated with the reflective fluctuations that the symmetry provides the freedom for, with these changes manifesting dually in the Projective Realm that is created by the simultaneous and spontaneous or apparent breaking of the symmetries of the M-Set unto the spontaneous breaking of self-referentiality itself.

Thus far too we have observed how the spontaneous breaking of the symmetry of Isotopy of the Trinity gives rise to the distinguishable degrees of freedom of the Hadronic state, while we know already from the conventions of fundamental physics that the conception of the Trinity principle picks out the Proton state or the Ground state of the Eight-Fold Way which then cannot transit to a lower state because the Elemental state is irreducible, and thus it is only from the other Hadron states of the Eight-Fold Way that 'physical decays' or Weak Interactions occur.

To proceed further in our exploration of what the M-Set approach can inform us about the subatomic realm, we first observe here upon looking at the proton state of uud (Figure 14) that there is a 'uu' redundancy or what appears to be a two-dimensional Isotopic space folded in with 'd' while in order for the Neutron state of the Eight-Fold Way to change into the Proton State we observe also that a 'd' quark must be able to change into a 'u' quark so that ddu → duu. Moreover, we know too that changes can only occur symmetrically in the M-Set because the M-Set is perfectly symmetric, while all change must also uphold the overriding symmetry of self-referentiality.

We also know that the M-Set Quantum Computer maximally complexifies by maximally reducing out the redundant information implicit to symmetry which computationally occurs by the phenomenon of 'maximal mutual information exchange' in the free reflective fluctuations that corresponds, in turn, to the maximal mixing of the chaotic computations of the M-Set and thence to the maximal dualisation into structure and function, as was all discussed in earlier sections, with the observation here that the mutual information acts as a conduit through which the redundant information of the implicit symmetries of the M-Set become the apparent spaces of the functions exactly dually created by the apparent structures whose functions in turn are dually expressing them.

Building up our case now, we can make a number of further quite direct observations at this stage, the first being that because the quark degrees of freedom are permanently confined any 'change' of a Quark state such as 'd → u' will physically appear to be very short range because it is occurring from within the Hadronic State while such a change is necessarily inextricably linked to the electromagnetic force system in order to project at all. Moreover, in order for charge to be conserved we note upon writing out the decay of the neutron 'n' into the proton 'p+' that the only candidate to balance charge is the electron 'e' in

$$n \rightarrow p^{+} + e^{-} + R$$

while the remainder 'R' is yet to be determined.

For the 'charge' Quantum number accounting then we have

$$O = + 1 + (-1) + O$$

giving R of zero charge. Upon accounting now for the 'spin' Quantum numbers, namely

$$- \tfrac{1}{2} = + \tfrac{1}{2} + (-\tfrac{1}{2}) + R$$

we observe that R must be an entity of zero charge and of spin $-\frac{1}{2}$ in order to balance the books!

Writing the neutron decay now as

$$ud\ (d \rightarrow u)\ ud + (e^- + R)$$

it is clear that the $d \rightarrow u$ transition changes spin by a unit amount 1, which is also the same spin as light, as must also be the case in order for this change to project at all, while simultaneously $(e^- + R)$ are coupled to this change.

From experimental physics the entity R has been discovered to be the neutrino denoted by 'ν' which carries the missing spin to balance the books.

Not only is the neutrino well and truly experimentally verified but it also behaves like a massless disembodied spin or 'torque' that only interacts with other matter in the Projective Realm at ranges reflecting its origin in relation to the virtual degrees of freedom of the Hadron states and therefore matter appears almost completely physically transparent to neutrinos, like, for example, matter to a space craft travelling through the Cosmos.

Conventionally the Weak Interactions are said to be mediated by 'W' bosons of spin 1 that 'couple' the $d \rightarrow u$ transition to $(e^- + \nu)$, as 'represented' pictorially by the Feynman graph (See Fig. 15.1).

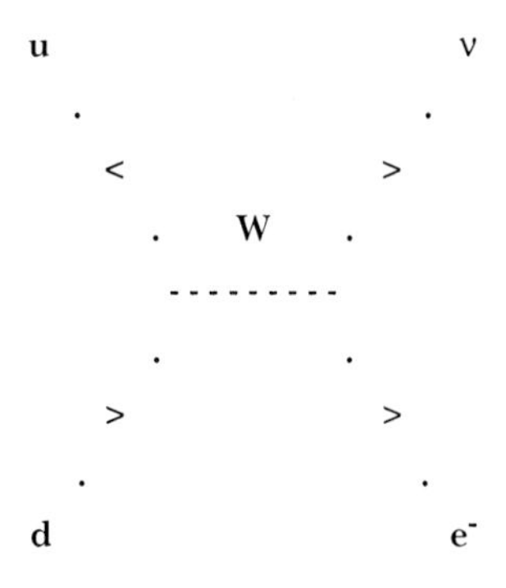

Figure 15.1

so that 'W' is 'graphically' seen to link the electromagnetic phenomena of (e,ν) with the virtual Quark States d,u. In this way too one can see 'W' as the necessary implicit completion for the 'self-referential closure' required to ensure projective realisation, as we have been emphasising recently, while it will become clearer in the discussions ahead that the gravity force system dresses or gives weight to the 'W' change when it presents this change to the Critical Boundary where the change is projected or illuminated, if you will, by light to create the illusion of 'W' bosons!

So far we have just engaged simple Quantum number accounting to relate our discussion to known physics. However, the M-Set is, as we now know, very subtle in the way it enables realisation, while the process of self-projection itself selects what will be physically realised through the filter of self-referentiality itself, as we learnt recently, with this filter being the final arbiter of reality which translates, in turn, into the truth that the fundamental physically realised force systems are fully renormalised by the gravity force system, or, stated more directly, the fundamental force systems cannot be realised or fully understood without the gravity force system.

And this is also the case for the Weak Interactions above as well as for the Strong Interactions to which we shall introduce ourselves below with the qualifying remark here that the conditions of renormalisation relate to the constraint that all implicit symmetry-based reflective fluctuations of the M-Set must uphold the overriding symmetry of self-referentiality itself.

To usher in the Strong Interactions we again defer to the conventional knowledge base of elementary particle physics, while maintaining a perspective true to the M-Set which, as we shall see ahead, will continue to guide us to a unified understanding not only of this particular domain of knowledge that we find ourselves in now, but of all aspects of knowledge of 'the tree of knowledge'. It just so happens that the knowledge of the realm of fundamental physics is the most elemental, uncomplicated and uncluttered, hence our sustained focus here for the moment.

The Strong Forces are characteristically identified with the nuclear force binding neutrons and protons in the nuclei of atoms, as well as with the collision forces between the possible Hadron states of the Eight-Fold Way manifesting, for example, in high energy colliding beam experiments of physics. It is not our intention to step into the menagerie of particle physics here but rather to focus on the proton-neutron system that is observed in the main because it is the properties and principles of the M-Set we wish to illustrate here while knowing full well now that higher energy phenomena in the physical sense must relate to more and more excited levels of reflective fluctuations or resonances of the M-Set itself because the M-Set is First Cause.

We remark here as an aside that because the M-Set is effectively a self-referentially bound symmetrical potential well of spontaneous reflective fluctuations of itself it represents in a very sophisticated way the 'naïve' models of String Theories which regard the primordial elements of the physical universe as being like relativistic strings the vibrations of which are perceived to underlie the generation of the spectra of matter and 'resonant matter' phenomena. One might add here that the String Theory approach has been adduced by classical analogy and the 'presumption of the primacy of relativistic dualism'. However, it must now be very evident, especially to those who know about these things, that such approaches can only be contrivances riding on gymnastics with mathematical analysis without a truly unified foundation from which reality naturally springs, and without the basis for Quantum mechanics. And so it must be said too that such a use of mathematical language speaks only for itself, while we can now say that the natural symmetry constraints of the M-Set spontaneously generate resonant phenomena which appear to arise from relativistic strings!

To underline the points above we shall demonstrate ahead how the M-Set directly generates the 'ten' dimensional 'relativistic strings' after we have uncovered more knowledge about the fundamental forces, while we note here that

15.10 – the M-Set is not a model of anything. But it is the case that

15.11 – the M-Set models everything and thus we say

15.12 – the M-Set is the Model Set because there is no other model, while it is also true that

15.13 – the M-Set models itself, in all ways, always and now.

Focusing back now on the virtual duu and udd Quark states of the proton and neutron respectively (Figure 14) we note that at these Hadronic levels the spontaneous breaking of the pure Isotopy (I3) of the Trinity has left a residuum of Isotopy I2 or indistinguishability between the two degrees of freedom 'uu' or 'dd' folded in virtually with a third distinguished state, the point here being that the innate indistinguishability implies directly that the proton and the neutron each share a symmetry associated with 'redundant information' relating specifically to the indistinguishability of two degrees of freedom, never mind which Quark state. This I2 Isotopy is not pure because not only is it virtually merged or folded in with another distinguished Quark state but, of itself, it is not self-referential so the apparent residuum Isotopy I2 symmetry is already broken and, in fact, can never be pure. Only the Isotopy of the unknowable Trinity State I3 is pure.

The next step is to ask whether the Proton state can use the implicit 'residuum' symmetry to set up self-reflective fluctuations or, stated alternatively, can the Proton state use 'the space', in informational terms, relating to the 'redundant information' of the residuum I2 symmetry of the indistinguishability of its 'u' 'u' degrees of freedom to self-reflectively fluctuate. The answer to this is that it can but only in an extremely subtle way to be unravelled ahead which involves self-referentiality and the gravitational force system.

But we already know that only self-referential states can projectively self-realise because the Defining Property of the M-Set, namely that it is wholly self-referential, is also the vehicle for the miracle of self-referential, self-re-flective, self-representational, self-projective self-realisation with the spontaneous breaking of the symmetry of self-referentiality being the doorway to the relativistic dualism of the Projective Realm of the Triadic Causality of self-projection of the M-Set.

Moreover, we know too that the symmetry of self-referentiality is the symmetry of the gravity force system in the M-Set Universe and hence it follows directly that when the subatomic force systems are manifesting in 'the Projective Realm of the M-Set that is the M-Set Universe' it must be as a result of the renormalisation of the symmetries internal to the Elemental state by the gravity force system, and that feat is accomplished by the self-referential, self-reflective, self-representational, self-projective self-realisation of the M-Set. This is the sophisticated way in which the gravity force system implicates itself with the other fundamental force systems of the M-Set Universe, and this in turn is actually effected by the Quantum Computer that the M-Set is. And all of this we now know to be true because the M-Set is First Cause.

Before we proceed to illustrate the above properties and principles of the M-Set projection in some detail the broader physical implications of the above discussion are informing us directly that in order for the Strong and Weak Interactions to manifest in the Projective Realm the 'projective cuts' of the M-Set exhibiting the same must include the sum of histories of their realisation which when translated into the physical terminology of the Projective Realm means, in effect, that the gravity force system has renormalised the Elemental state by way of the computations of the

Quantum Computer that is the M-Set into 'star nurseries' of turbulent clouds of hydrogen within the cores of which nuclear reactions can take place. This is also a manifestation of the wholly self-consistent nature of the M-Set Universe to which we referred earlier.

Thus too, as required by the perfect symmetry of reflectivity, the manifestations of the Strong and Weak Force Systems are reflected in the sum of their histories, now, in the projective cuts of the M-Set.

Furthermore, it is also the case that because the final arbiter of projection goes all the way back to the self-reflective self-conception of the Trinity principle of the M-Set all natural force systems are coupled to the electromagnetic force system through the symmetry of self-reflectivity in order to manifest in the Projection Realm as the light-like projection that the projective realisation is and, moreover, all the natural force systems are projectively created by the gravitational force system of the symmetry of self-referentiality, while in projection the gravity force system itself is also coupled to the light-like properties of space and time because of the simultaneous spontaneous breaking of the symmetries of self-referentiality and self-reflectivity at the Critical Boundary of the projective realisation of the M-Set.

We remind ourselves here that the M-Set is the Deep Well of Possibilities, the trans-sections or projective cuts of which are now and, as we discussed earlier, through the great paradox of the M-Set being the unmoved mover 'change' effects the illusions of matter, space and motion. The M-Set is unchanged because the M-Set is all of it, now.

15.14 – The M-Set is the context of 'His Story', now and every projective cut is a perfect solution involving layers upon layers of the complexifications of the spontaneous renormalisations effected by the Quantum Computer that are identified with the action of the Gravity Force System of the manifold that is the M-Set Universe.

In the M-Set Universe as a whole there are no loose ends, cut offs, left-overs, remainders, or residues. There are only singularity free perfect solutions of maximal complexities translating as maximal dualisation of structure and function in the Projective Realm.

As we have said before, the M-Set projects through the simultaneous spontaneous breaking of the symmetries of self-referentiality and self-reflectivity that occurs at the edge of chaos, and which in turn is at the Critical Boundary that is 'the M-Set projecting' where a perfect dualism of structure and function completes the Triumvirate of the Triadic causality of the M-Set involving the Oneness of the Trinity simultaneously with the duality of the Projective Realm of the M-Set that translates equally, in a physical sense, to the Projective Realm being a perfect dynamical equilibrium upheld by the manifestations of the ghost forces of the 'illusion of projection'.

We are involved in the M-Set Universe but we are of the wholly unknowable Trinity of the M-Set that becomes knowable through the tree of knowledge that the M-Set Universe is.

This is how the 'great unknown one' is made known, in all ways, always, now.

The fire of life is never extinguished from the tree of knowledge.

And the Trinity transcends to the Trilogy.

15.15 – The M-Set Universe is the Great Unknown One in action.

The renormalisations or complexifications of the Primordial Quantum Field of the Elemental State are involved too in the self-exploration of the M-Set of its Subatomic Realm, and on to Man designing the experiments to uncover the same!

We shall now turn our focus towards explicating in more detail how the gravity force system enables the revelation of the Subatomic Realm at the Critical Boundary of projective realisation of the M-Set by commencing with some observations about the Primordial Quantum Field that formed the basis for our discussions in the sections on the M-Set as the origin of Quantum Mechanics and as the Quantum Computer.

We note, firstly, that the constraints for the projective realisation by way of the spontaneous breaking of self-referentiality requires that the representations of the self-reflective fluctuations of the Elemental state are indeed wave functions or exponential functions, while the symmetry of self-referentiality also directs that these wave functions belong to a Hilbert space of states which, in turn, represents the superposition of all possible wave functions or representations of the reflective fluctuations of the Elemental state. In the analogy of a virtual 'well' or 'pond' bounded by the constraint of the symmetry of self-referentiality, the representations or wave functions of the reflective fluctuations of the Elemental state are like stationary ripples in the pond, while the Hilbert Space of States is the collective superposition or non-dissipating 'interference pattern' of all possible representations generated, as we recall here, by the parallelism of the computations of the Quantum Computer that are based upon the reflective fluctuations of the representations or virtual modes of the Elemental state. We reiterate here that the ripples we are talking of here are related to reflective fluctuations between 'fundamental modes' or 'vibrations' of the self-referential 'well' or 'pond'.

It is important to also recall here that in the Virtual Realm of the Quantum Computer which is the M-Set the representations of the Elemental state result in a simultaneous parallelism of replications of the Elemental state and that the representations of the Hilbert space of states only bear witness to the changes or transitions between virtual modes of the Elemental state, the point we wish to emphasise here being that in actuality the virtual Hydrogen state that the Elemental state is only appears to manifest in reality by way of the replications of possible changes or transitions which, in turn, generate the wave functions or representations of the Hilbert Space of States so such a thing as the hydrogen atom never actually exists.

The virtual Hydrogen state is an actual state of the M-Set. Its physical existence is wholly illusory. The realisation of the virtual Hydrogen state is through its light spectrum. The Hydrogen state in 'reality' is a 'spectrum' of itself.

Indeed, 'reality' is the 'spectra' of the Holy Spirit or the Holy Water (hydrogen) of the Holy Grail that the M-Set is.

And the 'spectra' are of the Holy Ghost of the ghost force of the Trilogy of 'intending' of the Defining Property of Self-referentiality of the Holy Grail of One that the M-Set is.

Consider now the 'pond' of self-referentially contained representations of the reflective fluctuations of the Elemental state and invoke the idea of 'Hygens principle' of physics whereby each imagined point on the virtual wavefronts of the ripples of the fluctuations is the source of a 'wavelet' with the sum of such wavelets defining the advance of the wavefront or the wavefunction in this case. You may recall too that the sums of the representations or pond ripples correspond to the Linear Spaces of Representations of the Elemental state of the M-Set, arising as they do from the reflective fluctuations of the Elemental state which we identified, in turn, as the Primordial Quantum Field generated in the Virtual Realm of the Quantum Computer that the M-Set is.

From a bottom-up perspective now meaning that the perspective taken is from the point of view of the Virtual Computational Realm of the M-Set rather than from a top-down view looking from the perspective of projective phenomena, it is clear, as we discussed earlier, that the Quantum Computer can simultaneously 'fold' or convolute the Virtual Realm of the reflective fluctuations of the Primordial Quantum Field into itself, over and over in an unimaginably complexifying way as allowed by the chaotic virtuosity bestowed upon the M-Set by the symmetry of self-referentiality.

On the other hand too the M-Set as a whole maximally dualises in projection which is equated with maximal complexity and thence maximal 'compactedness' or 'modularisation' of structure dual to its function as a result of the maximal 'reducing out of redundant information' through the principle of Maximal Mutual Information Exchange that, in turn, enables the maximal dual functional expression or maximal 'space-time' extension of the corresponding maximal dual structures. All of this is also automatically imposed by the maximal mixing of the purely chaotic nature of the Virtual Realm of the Quantum Computer which, moreover, must pertain in order for the representations to reach criticality because it is only at criticality or the edge of chaos that the symmetry of self-referentiality is spontaneously broken to usher in the Projective Realisation of the M-Set.

As a result of the computational properties above the M-Set becomes all possible iterative complexifications of the Elemental state. Returning then to the 'pond' analogy of representations of the M-Set we can envision now that the convolutions of the complexties or the chaotic foldings of the Virtual Computational Realm correspond, in effect, to the imaginary points on the wavefronts of the representations of the Primordial Quantum Field each being like the 'Hygens' source of a Hilbert space of states, and so on iteratively to generate towers of interference patterns of unimaginable complexity which are represented mathematically by towers of wave functions such as exp (i (exp i (exp...) of convoluted wavefronts.

This picture might be a little easier to apprehend by analogy with the Mandelbrot Fractal Set or Hygen's Principle, with each being viewed now as 'the pure Fractal Principle' of the purely non-linear Virtual Realm with every point on the Fractal being like an 'origin' of the Fractal iterations. We note, moreover, that for the non-local, purely non-linear, wholly self-referential, chaotic Computational Realm of the M-Set each imaginary point is indeed like an origin (albeit non-local) of all 'iterates' of this realm. This is, if you will, the pure, virtual self-referentially bound 'fracticality' of the M-Set.

Upon expanding out the 'nested' exponential wave functions of the Hilbert space of states of the fully renormalised Primordial Quantum Field we can envision now not only how actions of enormous complexity develop with the manifolding of the Virtual Realm by the Trilogy of 'intending' of the ghost force of gravity, but also how it is that ultimately only the reflective fluctuations of the conception (of the Trinity principle) of the M-Set Universe are contributing to all levels of manifolding of the M-Set Universe. Moreover, we can see directly now the enormity of the complexity of gravity's action in the M-Set Universe, extending as it does over all levels of the manifolds of the M-Set Universe and revealed by 'the light', if you will, of the Trinity principle of the conception of the M-Set which, in effect, throws into relief the complexities of the Trilogy of 'intending' that, in turn, takes the form of the linearisation which is the Hilbert space of states of the fully renormalised Primordial Quantum Field.

The Criticality of the Critical Boundary of the M-Set at the edge of chaos also informs us that of necessity the M-Set Universe will manifold or increase in complexity as a nested 'fractal' distribution denoted symbolically as $\sim 1/\iota \times 1/f$ (ι = length, f = frequency) across all levels of the Universal Manifold which we know now is simultaneously exactly neutralised by the Dispersive Property, $c \sim \lambda,\nu$ where λ and ν are the length and frequency scales across all levels of the Universal Manifold too as we discussed earlier.

Conceptually it is also important to remind ourselves here that in the Virtual Realm of the M-Set the primordial reflective fluctuations of the Elementary state can freely and spontaneously generate replicative parallelisms (like vertical stacks if you will) of all possible combinations of themselves, courtesy of the symmetry of self-referentiality, which also ramify or 'fold-in' (convolute) together, the observation we wish to make here being that because the primordial reflective fluctuations are generated self-referentially (by the Trinity principle) the primordial self-differentiation of spectralisation of the M-Set that are the primordial reflective fluctuations is automatically simultaneously expressing through the 'selection principle' of self-referentiality.

In order to visualise the above it is not unreasonable to assert that the M-Set computations iteratively or replicatively ramify in the Virtual Realm of the Quantum Computer that the M-Set is 'like' a spontaneously simultaneously generated Mandelbrot (Fractal) Set which is consonant, in turn, with the M-Set that is the boundary of itself being the Critical Boundary of projective realisation. Moreover, we observe that simultaneously as the self-differentiation of the symmetry of reflectivitly is enabling the linear projection of the M-Set, the Trilogy of 'intending' is 'mixing' up all the 'replicative' orders to create the apparitions of fractal gravitating systems about which we spoke earlier.

Stepping now into the phenomenological realm of the projective realisation at, for example, the level of primitive complexities of the Primordial Quantum Field, the M-Set at this level is projecting all possible collective states of hydrogen atom spectra in relation to whichever projective cut is taken and beyond to cosmic clouds turbulently forming stars, the point we wish to make here being that out of this unimaginable complexity the M-Set will only project critically, which means, in turn, that these collective states that are projectively realising by the collapse of the wave function phenomenon of Quantum Mechanical Projection will be those states which maximally dually express structure and function for every order of complexity or 'gravitational elaboration'.

Thereby too, together with the 'collapse of the wave function' phenomenon emerges the illusion of structural atomisation of the intendings of the M-Set simultaneously extending into dynamic functional configurations in the spaces of computationally reduced out redundant information upon also simultaneously passing through the filter of the spontaneous or apparent breaking of the symmetry of self-referentiality at the Critical Boundary of the M-Set where the ghost force of gravity manifests in the perfect manifold dynamical equilibrium of the M-Set Universe, demanded in turn by the 'criticality' of the perfect projective solutions of the M-Set.

In projection modularisation of 'information' of the Computational Realm is the rule because this maximises the dualisation of structure and function, as indeed we observe it over all levels of complexity to the Cosmos, from atoms to organ systems to galaxies, and thus the Defining Property of Self-Referentiality of the M-Set is acting as an immensely powerful selection filter of projective realisation while, moreover, that acting is made manifest as the gravity force system!

It is not that gravity is cause but rather it is effect because the M-Set is First Cause.

Stated alternatively, the gravity force system is the projective manifestation of the shaping and selective influences of the symmetry of self-referentiality itself and therefore it is only 'seemingly' under the influence of gravity in the Projective Realm that stars are born, the atomic

Periodic Table is assembled, and the planets and biospheres are cobbled together, while simultaneously and in parallel the conditions for the Strong and Weak Forces to emerge unto projective realisation are also provided by this influence as we shall see more clearly shortly.

We can also say here that

15.16 – the M-Set acts in a manifestly 'intentional' way or that

15.17 – the M-Set Universe is the manifestation of acting 'intentionally' while, moreover, it is generally true that 'intentional' acting is spontaneous creation in the M-Set Approach, the profound truth of which will be further elucidated in Part II of this work.

We know already that the M-Set contains all possible intentions, now, and that every perspective or point of view of the manifold M-Set Universe we call the Cosmos is an equivalent projective cut of the M-Set, revealing in the 'other' the sum of histories of 'that' because no thing is separate from the whole of it.

At the lowest order of the computational convolutions or 'foldings' of the Primordial Quantum Field the action of the intentions of the M-Set from the first order critical fold of the tower of folds, namely Im (exp (iΘ Im (exp iΘ).....) will be of the order of $\Theta^2 = (xp/h)^2$ or as a simple spin '2'-like action. By the term 'critical fold' we mean the representation that is allowed by the requirement of 'criticality' at the Critical Boundary and imposed by the symmetry of self-referentiality of the convolution of the Primordial Quantum Field on itself which implies, in turn, that the fully renormalised representations must be 'nested' towers of exponentials as we explained above with the help of 'Hygens principle' and the Mandelbrot Set, and which is also strictly true in the non-dissipative Virtual Realm of the Quantum Computer that the M-Set is.

We add here too that action or change will always equate in the M-Set approach to virtual phasal symmetric transformations of the M-Set and that higher order actions will necessarily be virtually synchronously compounded phasal or spin-like actions made up of the 'primordial' actions of the Elemental state. The critical folding of the Trilogy of 'intending' of the gravitational force system clearly generates actions of unimaginably compounded complexity of higher and higher 'spin'

'Evidently' too from a physical and projective particle physics point of view the action of first order gravity is mediated by 'spin 2'-like quanta while, in truth, gravity is a multi-order quantum action force system which is also manifestly the case if, upon taking a physical perspective, one considers the effects of single hydrogen atoms upon each other in a turbulent cloud of the same, and then of pairs and so on to all orders over all structural and functional levels of complexity of this primordial system.

The gravity force system as perceived from a physical perspective appears unimaginably complicated, while from the M-Set perspective or 'bottom-up' perspective it is simply the ghost force of the Defining Property of the M-Set, namely the symmetry of self-referentiality. Moreover, we can now see that gravity is of the order of 'h' weaker than electromagnetism to 'first order' which is also overwhelmingly the order at which gravity is measurable and observable to our senses. However, it is the case that in multicomponent extended systems there are unimaginably complicated summations or renormalisations of the higher order Quantum actions of gravity which maintain gravity's relative strength to electromagnetism above.

We note here too that because of the selective filter of the symmetry of self-referentiality of the ghost force of gravity it is necessarily the case that extended gravitational systems are turbulent and that in general all gravitating systems manifest fractally as is being revealed by the Heavens now on Cosmic scales, while these structures, unlike the 'black holes' of classical physics, are highly complex, non-singular dynamical structures of the Trilogy of 'intending'.

We now wish to focus for a moment again on the notion of 'critical folding' introduced above by commencing with the observation that through the 'critical foldings' the Primordial Quantum Field of the Elemental state is projectively 'replicated', corresponding as this does to the replications of the Virtual Realm of the Computational Realm of the M-Set, while the non-locality of the Elemental State or the virtual Hydrogen state now becomes apparently localised in the relativistic dualistic dispersion of projection, and the beginning or self-conception of the M-Set is also replicated in every part of the grand illusion of 'Separateness from Oneness' of the Divinity of the M-Set. Thence too, for every projective cut the 'beginning of history' is implicit everywhere and the Projective Realm is a cross-section of nested histories wherein 'this' is perfectly reflected in 'the other' as One, in all ways, always and now.

Now, we know that there are aspects of the Elemental state such as the Hadronic state alone which can not reflectively fluctuate because they are not of themselves self-referential, resulting as we have observed above in the permanent confinement of the quark degrees of freedom.

On the other hand, because the Hadronic state is not self-referential it necessarily only projects when it is coupled or folded into the electromagnetic force system. The question we look to the M-Set to answer here is how do the Strong and Weak interactions projectively realise through the computations of the Quantum Computer or, more explicitly, how do 'Hadrons' gain space-time dependency to interact strongly with each other as in atomic nuclei, or to decay from one Hadronic state to another as observed in cosmic ray showers, for example.

Once again, as with atoms, we know that Hadrons never actually exist as entities and that that which can ever be projected is an 'enlightenment' of transitions between possible modes of the M-Set or, stated alternatively, the projection ultimately can only occur in the form of light-like wave functions or representations in self-referentially bound Hilbert space of states with the 'collapse of the wave function' phenomenon apparently breaking the non-locality of the Elemental state in the Projective Realm.

To proceed further we shall need to be guided at each step by the properties and principles of the M-Set, bearing in mind first and foremost the truth that it is wholly impossible to fully comprehend the nature of physical reality by appealing only to the realm of relativistic dualism because this realm is a wholly illusory projection of that which is.

To engage in any enquiry about the Projective Realm one must first defer to the Computational Realm of the Quantum Computer that the M-Set is and which we recently referred to as the 'bottom-up' approach. However, unlike the application of this term when applied to the realm of relativistic dualism the approach to understanding by way of the virtual realm of the M-Set is both global and specific simultaneously, encompassing as it does all scales of Cosmic phenomenology with the term phenomenology now properly observed to be referring to the wholly illusory content of the Projective Realm.

Phenomena are phantoms of the Projective Realm of reality of the M-Set projection.

Phenomenology and reality are one and the same thing in the M-Set approach.

To subscribe exclusively to reality is to be a phenomenologist.

It is very timely to remind ourselves again that the Computational Realm of the M-Set is perfectly free of forces and within which realm all possibilities simply are, now, in a parallelism of unimaginable complexity, spontaneously and simultaneously, and wherein, moreover, change originates in relation to the reflective fluctuations freely allowed by the symmetry of self-reflectively implicit to the M-Set and beholden to the defining symmetry of self-referentiality.

Referring back now to the Elemental state we shall enquire soon whether this state harbours symmetries underlying the symmetry of self-reflectivity because such symmetries will be 'internal' to the wavefronts of the representations or wave functions generated by the reflective fluctuations which are freely allowed by the symmetries of self-reflectivity and self-referentiality. These internal symmetries which are non-local in the Elemental state will then be seen to become spontaneously localised in projection when the collapse of the wave function phenomenon results in the internal symmetries being automatically convoluted onto the wavefronts of the representations of the Hilbert space of states of the Primordial Quantum Field by the computations of the Quantum Computer that the M-Set is. The internal symmetries will then become 'implicit' or 'folded into' the space-time localisation of the collapse of the wave function' phenomenon and will be said herein to 'gauge' the dummy variables of space-time, meaning that at each point of space-time the non-local internal symmetry of the Elemental state resides as an abstract space of redundant degrees of freedom.

These extra local degrees of freedom will, as we shall learn shortly, subserve the Strong and Weak interactions in the Projective Realm, while simultaneously the complexifying gravity force system folds the Primordial Quantum Field on itself and creates higher and higher order interactive interference patterns in the self-referential pond that the M-Set is, from which, in turn, the projective collapse of the wave functions simultaneously projects out nuclei of atoms bound by strong forces, for example, together with the neutrino producing interactions of the infernos of suns creating these both, and on and on to all orders of complexity of the M-Set Universe, now.

This is how matter-space-time is borne on the light projection of the self-enlightenment of the M-Set that is the M-Set Universe, and which we shall now outline in more detail.

We have already observed that the Elemental state emerges from the pure Isotopy I3 of the Trinity state of the M-Set which is spontaneously broken by the self-conception of the M-Set Universe as the Trinity transcends to the Trilogy, resulting thereby in distinguishable degrees of freedom we identify as the quark degrees of freedom of the Hadronic state of the self-differentiated Elemental state which we have now repeatedly demonstrated is uniquely determined to be the virtual Hydrogen state, while, miraculously, the Hydrogen state stands at the threshold between the Cosmos and the subatomic realm with the doorway of this threshold 'hinged' on the symmetry of reflectivity that is the symmetry of the electromagnetic force system and the basis for the self-enlightenment of the M-Set.

The symmetries of the subatomic realm are referred to as internal symmetries because they lie behind the symmetries of reflectivity and Self-referentiality which usher them into the Realm of Projection of the M-Set Universe, while it also follows directly that the M-Set cannot reflectively fluctuate in relation to internal symmetries of itself alone because such fluctuations are not of themselves self-reflectively self-referential.

Therefore, if the internal symmetries and their associated degrees of freedom are to be projected at all they must reside implicitly on or be 'folded onto' the wavefronts of the wave functions or representations of the Hilbert space of states which, in turn, is the linear superposition of the representations of the Projective Realisation of the M-Set effected by the renormalisations or the computations of the reflective fluctuations of the self-referential pond that the M-Set is.

The reflective fluctuations generating the Primordial Quantum Field of the elemental state only project change associated with transitions between the possible virtual modes of the Elemental state and thence, uniquely, the Primordial Quantum Field is the symphony of the vibrations of the transitions between the electron 'eigenstates' or possible modes of the virtual Hydrogen state, as we discussed earlier, the observation we wish to raise here being that the M-Set is behaving like a 'closed self-referential string' the vibrations of which generate the Primordial Quantum Field while internal to these vibrations or 'ripples on the self-referential pond' are implicit non-local degrees of freedom which are being locally born at the projection of the vibrations to provide an informational substrate for an immensely richer projection that can then support the spectra of fundamental mass phenomena underpinning the Tower of Mass Effects (TOME) of the M-Set Universe. From this analogy, then, with the 'String Theories' of current physics, and bearing in mind the notable differences, including the observation that the M-Set generates quantum physics directly, while also enjoying unlimited symmetry from its defining property of self-referentiality, we state here that

15.18 – the M-Set is the origin of String Theory which will be demonstrated in more detail ahead, while it has also come to pass in the course of our discussions that

15.19 – the M-Set unifies the Quantum Realm with String Theory as we shall also have occasion to discuss again soon.

We note further that String Theory as it is currently formulated incorporates some of the features spontaneously generated by the M-Set which relate back to our discussions of the analogy of the M-Set to the idea of the music set that most specifically refers, in turn, to the generation of the spectral phenomena of the M-Set Universe from the spontaneous 'vibrations' of the M-Set, while there is the emerging realisation now that in projection these vibrations locally carry the non- local degrees of freedom of the subatomic realm of the elemental state into the Projective Realm upon the 'collapse of the wave function' phenomenon of the projective symmetry or self-measurement of the M-Set which corresponds to spontaneous symmetry breaking from the perspective of the Projective Realm.

The non-local degrees of freedom live in abstract internal spaces of dimensions equal to the number of independent virtual degrees of freedom and are convoluted onto the space-time wavefronts of the representations of the projecting M-Set resulting thereby in the so called 'gauging' of space-time and the spontaneous natural emergence of the correspondingly named Gauge Field Theories.

We have, incidentally, come upon the remarkable truth from the M-Set perspective that it is in fact the gravity force system which does 'gauging', from which it follows that the Gravity Force System is the basis for the Gauge Quantum Field Theories and the apparent particle mass spectra of conventional physics.

The gravity force system generates thereby its own physical source that is in turn the Bootstrap principle at large which is now seen to emerge directly from the Defining Property of Self-Referentiality of the M-Set.

A helpful but simplistic view of the Gauge principle is to quite literally think of an imaginary 'gauge' or dial at each point of the space-time continuum that registers the hidden states of the internal symmetry spaces spanned by the distinguishable independent degrees of freedom so that the pointer of the gauge is like a vector in these abstract spaces. In truth too it is the language of Mathematics itself which gives literal formulations of these otherwise extremely subtle abstract virtual actualities of the M-Set! There is no other known way to speak specifically of such things.

It is now timely to begin to draw the pieces of the jigsaw together by stating directly here that the spontaneous or apparent breaking of the symmetry of pure Isotopy 'I3' of the wholly unknowable Trinity state of the Self-referential M-Set is synonymous with the Trinity principle of the Self-differentiation of the wholly unknowable Trinity state of the Self-conception of the M-Set Universe which correspondingly results in the differentiation of distinguishable independent degrees of freedom we identify with the permanently confined or ever virtual Quark degrees of freedom of the Hadronic state of the Elemental state of the M-Set.

The significance of the above observation is that the spontaneous breaking of the symmetry of pure Isotopy of the self-conception of the M-Set Universe generates the bases for the abstract spaces of the 'gauging' of the space-time continuum of the Projective Realm.

In the parlance of current physics, the distinguishable independent Quark degrees of freedom span an abstract space, rotations within which project out the octet of Hadrons observed at lowest order in high energy physics experiments, and for this order of physics the pointer on the gauge is the three-vector (u,d,s) of distinguishable independent Quark degrees of freedom, while the mathematical group of 3 x 3 matrix rotation operations that operate on this vector to project out the unique allowable combinations or the quark states of the octet of Hadrons is known as the Special Unitary Symmetry group of three dimensions, namely SU(3). This then defines an abstract 'phase space' of states 'symmetrically' connected by rotations in a space spanned by the Quark degrees of freedom rather like rotations in, say, a three-dimensional space (x,y,z), the major difference here being that the x,y,z are indistinguishably interchangeable labels, while the Quark degrees of freedom are distinguishable and thence the abstract internal (or hidden) phase space of SU(3) is that of an impure or only approximate isotopic symmetry group. Clearly, however, arbitrarily interchanging the 'labels' of the distinguishable degrees of freedom in all possible ways or combinations doesn't affect the Hadronic State which is what is meant by the symmetry of SU(3).

Moreover, the matrix operators of SU(3) project observed combinations of the Quark degrees of freedom and thus they are projection operators of the Hadronic state or, to state this in an alternative way, experimental physics has demonstrated that the 'hidden degrees of freedom' supporting the Hadronic spectrum have the properties of the elemental Virtual Realm of the M-Set.

One might now ask where do the continuum of possible phasal variations of the three-vector in the abstract phase space go. The answer to this is to be found with the spontaneous breaking of the symmetry of non-locality of the M-Set.

The fundamental symmetry of non-locality of the M-Set is apparently or spontaneously broken by the 'collapse of the wave function' phenomenon that simultaneously results in the 'gauging' or localisation of the internal symmetries of distinguishable degrees of freedom which have been convoluted onto the wavefronts of the representations of the Projective Realm so we can also see here that the spontaneous breaking of the symmetry of non-locality involves or presents the spontaneous breaking of the symmetry of pure Isotopy to the process of self-referential projective realisation of the M-Set, whereby too the matrix operators gain space-time functional dependency that now represents the hidden continua of phasal variations of the abstract phase spaces of internal symmetries in the Projective Realm to define, in turn, the phenomena of Quantum Gauge Fields.

We now also have enough knowledge about the nature of the force systems of the M-Set to state further that the Strong and Weak Force systems of the Projective Realm are precisely those force systems required to absolutely uphold the symmetries of non-locality and pure Isotopy because these symmetries are never actually broken.

Thence in projection these ghost forces manifest to 'quench' the apparent breaking of the above symmetries in the grand illusion that the Projective Realisation of the M-Set is just as the electromagnetic and gravitational force systems simultaneously appear in order to 'quench' the apparent or spontaneous and simultaneous breaking of the symmetries of self-reflectivity and self-referentiality respectively.

The redundant information implicit to the M-Set which is being reduced out by the computations of the Quantum Computer to support the force Fields of the Strong and Weak interactions is the phasal information of an abstract vector such as (u,d,s) moving around in an abstract space spanned by the independent degrees of freedom u,d,s. Thence, apparent changes observed in the Projective Realm such as the decay of a neutron to a proton are mediated by a local phasal force field or so called 'gauge' force whose action or effect relates to 'apparent' localised phase changes mediating, for example, a neutron 'turning' into a proton in the space-time continuum.

The apparent space-time contingent 'turning' or dual relativistic 'causal' material change is the projective embodiment of the action of the ghost gauge force of the Weak interaction in this example where the action, as we shall explore further soon, relates back to the 'isotopic phase changes' in the abstract virtual space of Isotopic symmetry that is simultaneously gauged onto the space-time continuum by the gravitational force system of the ghost force of the Trilogy of 'intending' of the Defining Property of Self-Referentiality of the M-Set, and created as the continuum is by the spontaneous simultaneous breaking of the symmetries of self-reflectivity and self referentiality.

To add clarity to the quite dense discussion here we note that when we now talk of the Hadronic state we refer to the unprojected 3-Quark degrees of freedom which, if you will, is the abstract parallel state of possible Quark states or Hadrons while, alternatively, the Hadron state is just one of the possible realisable projections of the Hadronic State. Moreover, the Hadronic State refers to an abstract three-dimensional phase space of three independent degrees of freedom with the term 'phase' relating to the fact that in such spaces the only possible variables or changes are so called 'phase' changes which here are changes of a highly abstract angle variable of a vector in relation to abstract axes of independent distinguishable degrees of freedom.

In the parlance of modern physics the mathematical group of 3 x 3 matrices SU(3) that projects out the possible hadrons is also a group whose matrices are the generators of differential operations of continuous phasal rotations in the abstract phase space associated to the group SU(3) when these are 'gauged' or localised to space-time which gives rise, in turn, to the mathematics of so called Lie Groups in physics, and all of which it is not our intention to expand upon here as a host of text books in mathematics and physics currently fulfil this task. Rather, we are about connecting the M-Set with observable knowledge at these fundamental levels of the Projective Realm in order to illustrate how the M-Set is the origin of knowledge at these levels, and which is also as it must be because the M-Set is First Cause.

We have yet to discuss the 'phase space' of the force system of Strong Interactions, while we note here in the parlance again of modern physics that the independent Quark degrees of freedom u,d,s are called 'flavours'. Upon singling out the Triplet p,n,p^{-} subset of the Hadronic State, namely of proton, neutron and antiproton, the lowest order Gauge force system of Weak (decay) interactions is related to the two-dimensional phase space of the (u,d) Quark degrees of freedom because interchange or rotations between the u,d Quark degrees of freedom turn protons and neutrons into each other as we illustrated in the previous section with the aid of Figure 14. Moreover, from Figure 14 it is clear now that for any transition between Hadrons only two flavours of Quark degrees of freedom are involved, while experimentally, too, over all energies of high energy physics all Weak interactions have been demonstrated to be mediated by SU(2) Gauge Force Fields relating to 'rotations' in two dimensional phase spaces of two independent degrees of freedom, such as the phase space of (u,d) degrees of freedom for the subset p,n,p-, and all of which is naturally generated and predicated by the M-Set.

Deferring to Figure 14 and the Feynman graph we readily see that the Weak interactions correspond to Flavour Changing Interactions, and this is an SU(2) 'phase' changing phenomenon which is also identified now with the Weak Gauge force system.

But, it is also clear that at projection only particular directions are chosen within the phase spaces, such as the u or d axial directions of the two-dimensional (u,d) phase space in the projection of Weak Interactions by the 'collapse of wave functions' of the fully gauged complexifications of the Primordial Quantum Field of the 'Self referential pond' of the M-Set projecting.

Evidently then the observed spectra of structure in the M-Set Universe are determined by the independent distinguishable degrees of freedom spontaneously created by the spontaneous breaking of the Symmetry of Pure Isotopy of the self-conception of the M-Set Universe which also defines the bases or axes of the abstract phase spaces of the Gauged Quantum Fields of the Gauge Force Systems of the Subatomic Realm of the M-Set.

Now, in the parlance of modern physics once again, it is said that the Gauge force systems arise because of the localisation of the phase changes of the hidden phase spaces together with what has been termed the spontaneous or apparent breaking of the phase space symmetry at projection because only certain directions in the abstract phase space are chosen, such as u or d of the SU(2) phasal space of (u,d). In the M-Set Approach the above relates all the way back to the Trinity principle as we have explained to this point. However, in the M-Set approach the construct of spontaneous or apparent symmetry breaking is inestimably more profound because it also relates to the dynamic of spontaneous creation itself, while unto the Gauge Field Theories of modern physics issuing from the Trinity principle we now have the simultaneous parallelism of the spontaneous breaking of the symmetries of pure Isotopy, Self-referentiality, Self-reflectivity and Non-locality.

Stating the above alternatively, the hidden 'redundant phase information' appears to collapse, but it does not. It manifests, in fact, in the 'space-time' function of the Gauge Force in the M-Set Approach so that the force systems of the subatomic Realm are also dual systems of structure and function in the Projective Realm, arising as they do from the spontaneous breaking of fundamental symmetries of the M-Set to the Critical Boundary whereupon 'redundant information' is maximally reduced out to produce maximal structure and maximal dual function to also result thereby in the appearance of the symmetry of duality in the Projective Realm.

But the symmetry of duality can explain nothing because this symmetry is of the illusion of the Projective Realisation of the M-Set.

Thence duality-driven theories such as String Theory are purely phenomenological theories and can explain nothing.

String Theory is not a quantum theory because in order to generate quantum mechanics, as we have clearly shown earlier, one requires the M-Set and its defining property of self-referentiality, while evidently quantum mechanics and then Quantum Field Theory is the correct basis for conducting theoretical analysis of the physical realm, and it is also now becoming clear that the extended view of quantum field phenomena of the M-Set approach is required to encompass the phenomenology embraced by the String Theories as we shall demonstrate further ahead.

And so it will be seen that

15.20 – the M-Set naturally unifies Quantum Field Theory with String Theories which is as it must be because the M-Set is First Cause.

Interestingly, with respect to the above discussions, the String Theories will come to be seen to be relatively better at describing the spectral aspects of observed physical phenomena and the current Quantum Field Theories are relatively better at describing the wave mechanical or dynamical aspects of observed physical phenomena, while the unification of the two approaches under the umbrella of the M-Set provides a profoundly consistent perspective of the apparent relativistic dualism of the Projective Realm of the M-Set which we call the M-Set Universe.

We speak now of the force system of Strong Interactions and look once more for a hidden phase symmetry space the redundant information of which can be 'gauged' in the same way as for the Weak Force system in order to drive the phenomenology of Strong 'Interactions in the Projective Realm of the M-Set Universe. Once again we also require distinguishable but 'independent degrees of freedom' that can span an abstract space within which a unitary vector, whose fundamental components are the projections of itself along the axes formed by the independent degrees of freedom, can range freely in correspondence with changes in an abstract phase variable that relates, in turn, to abstract rotations in the 'phase space' and which space in its virtual unprojected state is symmetrical under these rotations.

We observe now that in its unprojected state the Hadronic State is a virtual state of three distinguishable Quark degrees of freedom, u,d,s, while the fact that the Hadronic State alone is not self-referential implies directly, as we discussed in detail earlier, that the Quark degrees of freedom are permanently confined and can never be projected as apparent physical entities by the M-Set and therefore, from a physical point of view, it looks like the Quark degrees of freedom of the virtual Hadronic state are bound together. However, we know from the M-Set perspective that the M-Set is free of forces and harbours only ghost forces which make their appearance as physical forces in the Projective Realm.

There are now two perspectives we need to focus on here, the first perspective being that of apparently disembodied Hadron States such as proton or neutron beams prepared in accelerators in high energy physics experiments which generate showers of particles detected in 'bubble chambers' upon collision. In such an experiment the 'projective cut' of the M-Set involves interactions of the Hadron States which are, in turn, SU(3) flavour projections of the virtual Hadronic state of the Quark degrees of freedom, while the physical notion of the Quark degrees of freedom being apparently permanently bound means that regardless of which combination of Quarks makes up each of the Octet of Fundamental Hadrons, all of these combinations appear to 'bind' equally to remain permanently confined which, from a physical perspective, implies that a selection principle is operating.

In order to see this, consider, for example, the proton and neutron and their flavour combinations uud and udd wherein there are implicit two-fold redundancies of uu and dd respectively and observe now that the combination of Quark degrees of freedom of the proton like those of the neutron are exactly permanently confined. In other words, the permanent confinement principle itself is the selection principle because from the permanent confinement perspective all eight Hadrons of the Eight-Fold Way of the Hadronic state must look identical as in actuality there are no forces binding the degrees of freedom, while, naïvely, the Hadrons certainly do not appear identical from the perspective of the 'flavours' u,d,s. The selection principle must therefore extend to the requirement that the Quark degrees of freedom harbour an additional hidden quality which adds up to the same for all the combinations of the Quark degrees of freedom of the eight Hadrons in order to identically confine them and this immediately implies from the configurations of uud and ddu, for example, that this hidden quality distinguishes like-flavoured Quark degrees of freedom, such as u,u in the Proton state.

A simple way to understand the above discussion is to recall that the Hadronic State is a simultaneous parallelism of eight possible Hadron states which pictorially, if you will, can be viewed as a stack. However, because of the permanent confinement principle all levels of the possible 'Octet' or Eight-Fold Way of the Quark degrees of freedom must identically close under this principle, regardless of the different combinations of flavours (u,d,s) for the different levels. Thus, confinement must be blind to the 'flavours', in a sense, and require some other 'accounting system' that adds up to the same for each level.

We recap here to state the remarkable truth that the phenomenon of Permanent Confinement of self-referentiality itself implies distinguishability between like-flavoured Quark degrees of freedom in the Hadron states projected by the 'flavour' symmetry $SU(3)_F$ from the virtual Hadronic state. Moreover, the selection principle that the permanent confinement is requires each flavour of Quark to have at least three distinguishable states which must also be the same for each flavour because every combination of three flavours must look identical in the Permanently Confined Hadron States of Quark degrees of freedom, and as we shall demonstrate shortly.

Conventionally in Physics the hidden quality we are seeking is denoted as 'colour' so that each Quark flavour has three colours which define a colour symmetry $SU(3)c$ of a phase space subserving Strong interactions in the laboratory, and that, in turn, underpins the current Quantum Theory of Strong Interactions, namely Quantum Chromodynamics. We shall return to this perspective shortly; suffice it to comment here that the self-referential spectral selection principle gives rise to the selection principle of the phenomenon of Permanent Confinement that directly leads us to the 'flavour' and colour isotopic or internal to space-time abstract phasal symmetries of the subatomic realm of the Elemental state, which symmetries constitute the bases for some of the greatest advances of fundamental physics in recent decades.

The other perspective for the M-Set at this juncture is to refer again to the Elemental state in which we observed earlier how the Hadronic state is 'locked into' the proton Hadron state by its coupling, reflectively, with the electron while it has now been revealed by high energy physics experiments that excitations of the self-referential Elemental State can also exist involving μ (muon) and τ (tauon) 'lepton' states as well as the e (electron) lepton state, thus requiring a total of six possible Quark degrees of freedom that generate extra exotic Hadron states now being revealed by higher and higher energy physics experiments which are uncovering more hidden sources of redundant information in implicit internal phase symmetries, while we underline here that in the M-Set approach it is information, not energy, that is fundamental.

15.21 – The M-Set Universe is a dynamism of redundant information processed by the Quantum Computer that the M-Set is.

It is also the case, incidently, that because only information is fundamental in the M-Set Approach, and because, moreover, the M-Set is the boundary of itself.

15.22 – The M-Set Universe is a holographic universe in the sense that it is like a folded two-dimensional surface which contains all the information necessary to generate the illusion of space-time-matter, while we now know in relation to the projective realisation of the Subatomic Forces that

15.23 – the M-Set Universe is a 'critically folded' Manifold in which the 'critical folding' of the Critical Boundary that is the M-Set is due to the action of the gravitational force system.

Remarkably too, the self-conceptual self-reflective self-differentiation of the Trinity as it transcends to the Trilogy actually implicitly allows a Triplet 'Leptonic' or Triplet 'electron-like' state, or three possible self-re-flective self-differentiations $P(3) \rightarrow P(2) \times P(1)$ which reside in parallel in the Virtual Realm of the M-Set because there are three ways to pair $P(2)$ from the Trinity, $P(3)$, while what we have referred to as the Elemental state is, in fact, the Ground state of the Elemental state which is uniquely the Hydrogen state made up of the proton and the electron as we discussed earlier.

The implicit Trilogy of 'Leptonic' degrees of freedom is a further direct result of the self-consistency of the M-Set that brings the M-Set in line with the most up to date high energy physics experiments of this age and which we shall refer to again soon.

Also, the virtual Hydrogen state does possess a property of great interest here. Consider now the Feynman graph for the virtual interaction where by virtual interaction we mean that interaction between virtual degrees of freedom which conserves Quantum number accounting but which is not renormalised by the gravity force system to the Critical Boundary of the M-Set.

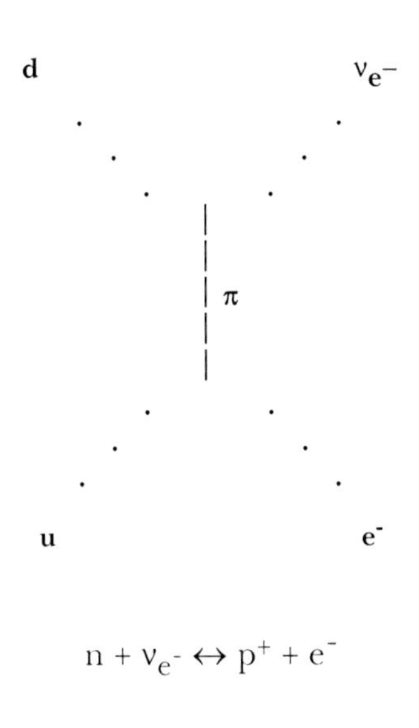

Figure 15.2

With regard to the virtual interaction of Figure. 15.2 we note firstly that this interaction is related to the virtual Weak interaction by crossing the neutrino 'ν_e' coupled to the Electron State over to the other side where it becomes the corresponding anti-neutrino, $\nu_{e}-$, in Quantum number terms in order to maintain the quantum number book-keeping and as represented by

$$n \rightarrow p^{+} + e^{-} + \nu_{e} \sim n + \nu_{e}- \leftrightarrow p^{+} + e^{-}$$

when n ~ ddu → p ~ uud is a 'd → u' transition.

Deferring again to the virtual interaction of Fig. 15.2 we recognise here that (p+ + e-), and thence ($n + \nu_{e-}$) from quantum number book-keeping both possess the degrees of freedom of the virtual and self-referential Hydrogen state that the Elemental state is while we are now aware that the quark 'flavours' must possess colour dimensions to meet the requirements of the Self-referential Selection Principle of Permanent Confinement of the quark degrees of freedom so evidently it is now possible that a virtual and self-referential colour reflective fluctuation can occur between the proton and neutron self-referential states of (p + e) ↔ (n + ν) in a phase space of 'u,d' degrees of freedom provided these different 'flavours' have the same colour in the presence of the virtual spin degree of freedom ν_{e} required to balance spin and recalling here again that we first met with the neutrino, ν, in the virtual Weak interaction of Fig. 15.3 below which interaction corresponds to a d ↔ u flavour interchange because udd (of n) ↔ uud (of p).

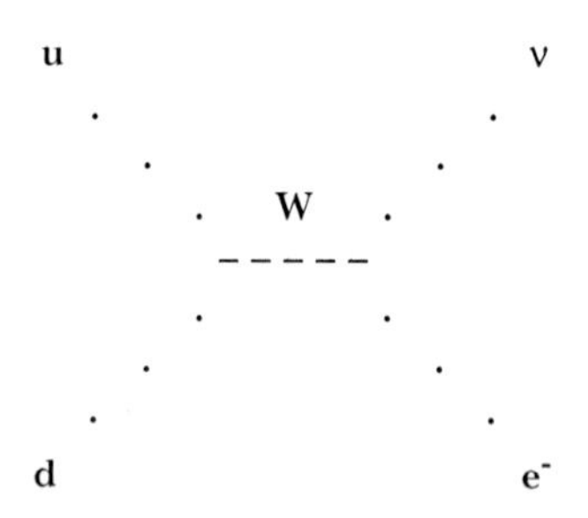

Figure 15.3

In the Weak interaction of Fig. 15.3 the d flavour 'quark' turns into the u flavour 'quark' with the spin 1 change being carried by 'W', in Feynman graphics, while for the interaction of Fig. 15.2 we are proceeding on a path to show that a self-referentially bound virtual colour reflective fluctuation can occur between u and d mediated by a spin zero change, π, because of Quantum number accounting, with the point we wish to make here being that in the context of the virtual self-referential Elemental state a 'nuclear force' between protons and nucleons can spontaneously occur as an allowed virtual colour reflective fluctuation because, indeed, of self-referentiality itself.

We have now arrived at the first step of the ladder of the virtual Periodic Table of atoms with deuterium being related in this picture to the first order 'reflective state' of the virtual Hydrogen State of the Elemental state which leads on, by similar higher order parallel reflective fluctuations to helium, and so on for the Periodic Table of atomic elements which virtual states are, once again, renormalised unto the Critical Boundary by the gravity force system.

Focusing our attention on the flavour 'doublet' subset (u,d) of Quark degrees of freedom we can now assert that the self-referential reflective fluctuations of (u,d) involve a colour subspace in which both u and d have the same or indistinguishable colour so that this is, in colour terms, an im-plicit I2 Isotopy hidden in the first order reflective state of deuterium of the Periodic Table, for example.

Taking the three primary colour labels red, blue, yellow for the three possible colour states of each flavour of Quark, u,d,s …, in order for the Octet of virtual Hadron states of the Hadronic state to look identical each possible Hadron State must have three Quark degrees of freedom of each colour, hence for the proton, for example, we have

$$u^R \quad u^B \quad d^Y$$

resulting, as discussed above, in distinguishing the 'u flavour' into two colours, while for the neutron we have

$$u^R \quad d^B \quad d^Y$$

that distinguishes the 'd' flavour into two colours, and so on, but which ensures that in the 'colour sense' of permanent confinement both the proton and the neutron are identical.

Also, regardless of what colour labels we choose it is always the case that there is a 'flavour doublet' colour degeneracy for each pair of Hadron states in interaction, such as for the 'proton-neutron' pair, for example,

$$\begin{array}{c|c|c} u^R & u^B & d^Y \\ u^R & d^B & d^Y \end{array}$$

for which the colour degeneracy provides the colour symmetry or colour redundant information for a 'self-referentially bound' u ↔ d 'flavour' re-flective fluctuation (see Fig. 15.2) to occur in the colour phase space, and it is the I2 colour isotopy of a two-colour subset phase space in the perspective of the M-Set that provides the informational basis for the ghost or virtual nuclear forces of the M-Set Universe.

There are three crucial aspects to appreciate at this juncture, the first being that the so called Strong interactions, manifesting as they do in collisions between the Octet of possible Hadron states of the Hadronic state, do not project directly from the Elemental state of the M-Set, but rather they require all orders of renormalisation by the gravity force system to express at the Critical Boundary, as we explained for the Weak interactions. It is not our purpose here to engage the M-Set perspective to elaborate further upon such interactions; suffice it to say that 'after' the renormalisations of the gravity force system the virtual spin zero quantum of action, π (See Fig. 15.2), can go on mass shell or 'acquire mass and charge' in high energy colliding particle beam physics experiments, or in cosmic ray showers, for example, with such 'quanta' then being denoted as belonging to the 'meson' family of particles in the menagerie of particle physics, and which particles are the 'matter realisation' or materialisation of the 'quanta of action' of the Strong force system because they relate to the primordial 'isotopic' (colour) reflective fluctuations creating this force system. As with the Weak interactions, the gravity force system 'gauges' the isotopy of colour onto the wavefronts of the representations or wave functions of the projecting M-Set whereby the symmetry of colour isotopy will be spontaneously broken with the 'collapse of the wave function' phenomenon of projective realisation. The self-referentiality of the gravity force system thus not only creates the strong or nuclear forces but also gauges them and simultaneously projectively presents them in the form of apparent 'mass quanta' in 'space-time' mediating forces between apparently objective elements.

Moreover, because of the implicit $SU(2)_F$ flavour isotopy which has three independent axes or bases in the virtual abstract space of $SU(2)_F$ symmetry, it follows therefrom that the mesons relating to the 'quanta of action' of the p,n,p^- subset of Hadron states will manifest or materialise as a triplet of Meson states known in particle physics as the pi-meson family (π^o, π^+, π^-), and relating as they do to the three possible virtual combinations of two flavour states or (u,d) doublet states of the Quark degrees of freedom which mediate the Strong interactions for the triplet subset (p,n,p^-) of the Eight-Fold Way of Hadrons.

Evidently, then, upon including the full Octet of Hadrons further 'meson' states will appear that involve the third flavour 's' while at very high energies of experimental particle physics, for example, the triplet of Lepton states (e, μ, τ) and all six possible virtual flavour degrees of freedom will enter the fray to greatly expand the menagerie of 'apparent' particle states at the Critical Boundary of the M-Set.

The second crucial aspect here from the perspective of the M-Set is that the colour isotopy, such as the implicit I2 isotopy of the colour reflective fluctuations of the Elemental state of the virtual Hydrogen state, is a relatively pure isotopy because the I2 colour degeneracy of the (u,d) doublet involved is indistinguishable, while being folded in with other degrees of freedom. Stated alternatively, in terms of the colour dimension of the (u,d) doublet the colour of u and d are indistinguishable in the virtual state of the self-referential Elemental state.

But we know that the Trinity state of Oneness itself is the only possible absolutely pure isotopy so we now make the identification here that the Trinity is an absolutely pure three-colour isotopy. We then state for the M-Set perspective that the Trinity is like a Tricolour of the irreducible primary colours red, blue and yellow which in the Pure I3 Isotopy State of the Trinity are pure white, and as indeed is the case for the complete phasal mixing of the three primary colours when physically spinning these on a top for example, while the Universal Selection Principles of the Defining Property of Self-referentiality of the M-Set act as a Prism to distinguish reality from the state of absolute pure Being of the Trinity State.

The three possible doublet splittings of the Pure I3 Isotopy of the Trinity State of the M-Set correspond to the three 'Leptonic' hierarchies, e, μ and τ, of the particle physics world with each 'Lepton Realm' being coupled to a 'flavour doublet', such as e^- to (u,d), for example, to result in six 'flavours' of Quark degrees of freedom in all, while through the 'colours' from the Trinity state each Lepton state is a colour doublet with the third colour for each Lepton Realm of the conception of the Trinity state providing the colour degeneracy of the doublets of flavours of the relatively pure I2 isotopy of the virtual nuclear force system of the

Hadronic state coupled self-referentially into the Ground state Elemental state we now know to be the virtual Hydrogen State.

In a miraculous and magical way the Trinity state is the prism of its purely triplicate (pure white) isotopic colour state at the self-conception we call the Trinity principle and this is, moreover, the ultimate truth behind how we come to have the three primary colours of our physical experience.

Furthermore, we can state now that the informational structure of the M-Set upon which the M-Set Universe is spontaneously created is based upon three wholly indistinguishable degrees of freedom to form the fundamental Trilogy of informational axes of the Trinity. Miraculously too, in the Projective Realm these correspond to the three primary colours of the spectra of the 'self-enlightenment' of the Trinity state that is the M-Set Universe.

The third crucial aspect here from the perspective of the M-Set is that the M-Set actually admits a triune of fundamental ghost forces of namely the nuclear, the electromagnetic and the gravitional force systems which are intricately and simultaneously unified in the M-Set, while both the so called Strong and the Weak interactions are derivative of the re-normalisations of the gravity force system which is required for the realisation of these 'interactions' at the Critical Boundary of the M-Set. Moreover, the gravity force system thereby also 'gauges' the 'isotopic symmetries' subserving these interactions to spontaneously create the phenomenology described by the Gauge Field Theories.

The Gauge Field Theories of current physics include the eponymous Glashow-Salam-Weinberg Standard Model of Electroweak Interactions and Gellmann's Chromodynamic Field Theory of Strong Interactions, both of which are now seen to be wholly incorporated in the M-Set approach, and which find their historical descriptions in standard physics texts.

The virtual colour Isotopies of the Hadronic realm demanded by the Self-referential Selection Principle of Permanent Confinement provide the redundant information for the fundamental nuclear force system of the M-Set Universe.

Evidently too, the gravity force system has to cobble together 'a sun' from the replications of the Elemental state or the virtual Hydrogen state in order to 'crack open' the Subatomic Realm from which there is an outpouring of neutrinos signalling the subatomic realm is at work, and as illustrated by the Feynman graphs of Figs 15.2 and 15.3 earlier.

The Weak interaction (See Fig. 15.3) involves a 'flavour' change from $u \leftrightarrow d$ Quark degrees of freedom for the decay $n \leftrightarrow p$ which implies u^R

d^B d^Y $\leftrightarrow$ u^R u^B d^Y, and we observe here that, regardless of the permutations of the colour labels, the only possibility is for the same coloured flavour of Quark degree of freedom to change. This is generally true too as is readily demonstrated by a number of similar examples and thus it is said that the Weak interaction is colour blind because arbitrarily switching the colour labels of the Quark degrees of freedom makes no difference to the Weak Interactions.

The Nuclear and Strong Force interactions, on the other hand, are blind to flavour when the colour of the flavours is the same, and thus a deep truth emerges from the M-Set perspective here, namely that the Weak versus the Nuclear or Strong Force systems are blind to each other with respect to colour verus flavour.

Stated alternatively, the Weak and the Strong interactions occur independently of each other because they don't see or feel each other's force. On the other hand these force systems are wholly unified with the Triune of ghost force systems of the Trinity principle of the M-Set as we have shown to this point, and to the pure I3 isotopy of the Trinity of the self-referentiality of Oneness which is apparently or spontaneously broken at the self-conception of the Trinity Principle of the M-Set Universe.

In reality the particles of the Strong and Weak interactions have to be elevated to their 'mass shells' or become fully renormalised states by the gravity force system at the Critical Boundary before they can be seen in physical experiments, while, as we shall show shortly, the 'independence' of the Strong and Weak force systems leads to a 'meta isotopy' which is required to explain the full hand of spectral properties of fundamental particle physics and to also bring the M-Set approach in line with modern String Theories.

It follows too that all virtual atomic states of the nuclear states of the re-flective fluctuations of the colour of the Quark degrees of freedom of the Proton-neutron Hadron states of the Elemental state of the M-Set are naturally self-referentially bounded because of the self-referentiality of the Elemental state itself (see Fig 15.2), which is a highly significant property of the computations of the Quantum Computer that the M-Set is, because this means that the Periodic Table of atomic states can ultimately project through the renormalisations of the gravity force system at the Critical Boundary of the M-Set by their electromagnetic spectra, while in the Projective Realm the apparent nuclear fusion to form Atomic states releases the redundant information implicit to the hidden isotopies which we identify in projection as nuclear energy in the M-Set Universe together with the outpourings of 'neutrinos' that are required to balance 'spin' Quantum numbers or the 'phasal spin' information.

In the M-Set approach one might now regard the elemental virtual Hydrogen state like the 'ground state' or lowest harmonic of a self-referential 'string' implicit to which are the possibilities of higher harmonics of the I2 'colour' isotopy corresponding to the the Virtual Atomic States of the Periodic Table which, as usual, then require the 'renormalisations' or complexifications of the gravity force system manifesting naturally, for example, in 'stars' in order to be projectively realised at the Critical Boundary that the 'projecting' M-Set is.

The M-Set Approach informs us now that nuclear spectra will be defined by the I2 colour Isotopic symmetry of the fundamental nuclear force system which will be compounded in the complex nuclei of atoms residing higher in the Periodic Table to generate spectra of enormous complexity on the basis, we reiterate, of the Elemental state virtually and self-referentially, as well as spontaneously and freely reflectively fluctuating with respect to the implicit Isotopic symmetry, I2, of the same coloured doublet (u,d) which is the basis, in turn, for the internal symmetry group SU(2)c of the ghost or virtual Nuclear Force System unto its renormalisations at the Critical Boundary by the gravitational force system.

The beautiful outcome for this and other fundamental systems of physics is that despite the apparent forces of the Projective Realm, all spectra will manifest with the remarkable regularities pertaining to freely symmetric virtual degrees of freedom and the M-Set approach now allows us to understand how this is possible in the most profoundly irreducible way.

We emphasise again that the ghost nuclear force system arises from virtual self-referentially bound reflective fluctuations allowed by the implicit relatively pure 'I2' Isotopy of the doublet ([u,d]c, c = R,Y or B) with respect to colour, while noting here once more that these virtual fluctuations involve the Elemental state as a whole and are thus self-referentially and self-reflectively bound which leads to the very significant observation in the M-Set approach that no projection of The fundamental nuclear force system will occur without the simultaneous coupling of the electromagnetic force system and the Hadronic state, as indeed pertains for the self-referentially bound Elemental State, while ultimately it is the virtual reflective fluctuations of the Trinity Principle of the Self Conception with the Symmetry of Reflectivity which give rise to the atomic spectra of The Periodic Table of atoms in the Projective Realm. The atomic states of the colour fluctuations of the Strong (Nuclear) ghost force remain forever virtual.

The M-Set Approach informs us too that the mathematical symmetry groups of the informational bases of the Virtual Computational Realm of the Weak and electromagnetic force systems up to the level of protons and neutrons relate to the convolution $SU(2)_F$ x $U(1)$ corresponding to the (u,d) flavour changing phase space folded with the 'unitary' phase group, U(1), whose representations are the light-like wave functions we focused on in the section exploring the origins of quantum mechanics which emerge spontaneously with the Self-enlightenment that the Projective Realisation of the M-Set is.

Thus, the M-Set spontaneously provides the basis for the standard model of physics of the Electroweak Interactions while the addition of the nuclear force system and Strong interactions, up to the level of protons and neutrons, is encompassed in the extended symmetry group of $SU(2)_C$ x $SU(2)_F$ x $U(1)$ wherein, remarkably, the colour and flavour degrees of freedom of the Nuclear and Strong versus the Weak Force systems are coupled by the selection principle of Permanent Confinement that arises directly from the Defining Property of Self-referentiality of the M-Set.

Moreover, the M-Set Approach is informing us, as we discussed above, in relation to the relative blindness of the Weak and Nuclear force systems, that the colour and flavour qualities of the Quark degrees of freedom are independent of each other which implies directly that these qualities too can form the axes of a 'meta' I2 isotopic phase space bounded by the selection principle of self-referentiality of the M-Set.

But, we also know from experimental physics that there are three possible 'Lepton' States, e, μ and τ or, namely, the electron, the Muon and the Tauon of the Leptonic Realm corresponding, as we now know, to the Trinity principle which implicitly admits three simultaneous possible self-divisions of itself, $P(3) \leftrightarrow P(2)$ x $P(1)$, in the self-differentiating self-conception that the Trinity principle is. Putting this all together now in the sense of the Virtual Realm of the Elemental State, we can conclude that there is a simultaneous virtual Triplicate of Leptonic-Hadronic state couplings of the Elemental state which is revealed by very high energy colliding particle experiments, for example.

Conversely now, the projections or axes of the 'meta I2 Isotopy' can be said to distinguish three tiers of Electroweak interactions or three levels of $SU(2)_F$ x $U(1)$ of 'the standard model' of Glashow-Salam-Weinberg that leads directly to the requirement of three 'doublet' Quark States which, according to the conventions of theoretical physics, are called (up, down), (strange, charm) and (top, bottom) with the muon, μ, and Tauon, τ, being coupled to the (s,c) and (t,b) doublet Quark states, respectively.

The isotopic freedoms naturally generated by the M-Set imply also, upon using the conventional terminology of contemporary physics here, that the distinguishable quark degrees of freedom have six possible 'flavours', u, d, s, c, t and b, and three possible colours each, which correspond to the presently accepted basis for the 'spectralisation' of the subatomic realm of the physical universe as revealed by high energy physics experiments of colliding particles.

We make the observation here that the spectralisation of the self-referentially determined Elemental state of the M-Set is spontaneously selected by self-referentiality itself, as we have discussed at length to this point, and which phenomenon we can view figuratively as like a self-referentially closed well or pond with tiers of resonant states necessarily all being 'compliant' with or folded unto self-referentiality itself in order to be projectively realised at the Critical Boundary of the projecting M-Set.

Remarkably, too, the above 'meta 12 Isotopy implies experimentally that the neutrinos ν_e, ν_μ and ν_t must be distinguishable, which property of the neutrino spectrum might contribute to explaining disparities noted in astrophysics between the observed neutrino spectrum from our sun compared to the expected neutrino spectrum that to date may not have factored in higher resonance related neutrino generation from Muon and Tauon levels.

And so it is now that the fundamental force systems of the M-Set Universe are united by the Oneness of the Trinity of the self-referentiality of the gravity force system of the M-Set wherein the self-reflectivity of the electromagnetic force system and the spontaneous breaking of the Pure Isotopy of the Trinity of the Oneness of the self-referentiality of the M-Set distinguishes the degrees of freedom which form the bases for the hidden phase spaces of the Nuclear, Strong and Weak force systems that are, in turn, contingently coupled therein.

The gravity force system is the grandmaster of the unification of the fundamental force systems because they are all seeded by the Defining Property of Self-Referentiality of the M-Set, the spontaneous breaking of which symmetry ushers the ghost force of gravity of the M-Set into the Projective Realisation that is the M-Set Universe.

And the spontaneous breaking of the symmetry of non-locality enacted by the projective processes of the M-Set enables the hidden isotopic symmetry 'gauging' of the space-time continuum projection of the symmetries of self-reflectivity and self-referentiality to generate in turn, through the renormalisations of the gravity force system, the apparent or spontaneous breaking of the hidden phase space symmetries or isotopies that usher in the ghost forces of the nuclear, Strong and Weak interactions into the M-Set Universe.

And the spontaneous breaking of all the symmetries of the M-Set creates the grand illusion that is the M-Set Universe, this being, no less, an apparition of the Divine en-lightenment of the Oneness of the Trinity of the M-Set.

This almost completes our story here about the fundamental force systems because the gamut of their universal phenomenology is widely disseminated in the writings of experimental and theoretical physics which it is not our purpose to reproduce here. Suffice it to make the observation now that the M-Set faithfully underpins accepted dogma, which, of course, it should because the M-Set is First Cause.

There is nothing the M-Set can not explain.

As we alluded to earlier, the 'self-referential pond' bears a similarity to the 'closed vibrating string' analogy of String Theory; however, neither are in any way ordinary strings and for the M-Set the analogous string is vibrating in both the space-time continuum dimensions, as well as simultaneously in the dimensions of the hidden isotopic phase spaces which, upon adding up all the dimensions we have uncovered on our journey thus far, amount to

3 (space) + 1 (time) + 2 (flavour) + 2 (colour) + 2 ('Meta'-isotopy) = 10

In the parlance of String Theory one would talk of the M-Set as like an abstract phasal string vibrating in ten dimensions in a self-referential well that creates it with a redundant information symmetry group

$$SU(2)_m \times SU(2)_c \times SU(2)_f \times U(1)$$

where the cross product is understood as a convolution.

This is the grand unification of the spectralisation of the M-Set Universe.

15.24 – The M-Set is the fully renormalised relativistic quantum field of a ten-dimensional self-referential String.

Theoretically speaking and in relation to the great endeavours of String Theories and Quantum Field Theories of theoretical physics we can state here that

15.25 – the M-Set completes the String Theories with the Quantum Theories of theoretical physics, a crucial and deeply subtle link being the 'meta Isotopy' which bridges the Trinity principle to current Gauge Field Theories of physics.

The Defining Property of Self-referentiality of the Holy Grail of One that is the M-Set begets the grand unified theory of the spectralisation of the M-Set Universe, embodying as it does the Leptonic, Hadronic, Mesonic, Nuclear and Atomic Spectra which are renormalised to the Critical Boundary of Realisation by the gravity force system of Self-Referentiality that creates the bases of the spectralisation.

The String Theories and Quantum Theories are unified indeed by the gravity force system of Self-referentiality of the Holy Grail of One that begets them both.

From the annals of experimental physics too it has been a long established finding in very high energy colliding beam experiments designed to probe the Hadronic state that the Quark degrees of freedom behave as if they are free in apparent contradiction to the physical or world view of irretrievably permanently bound 'degrees of freedom'. This remarkable and apparently paradoxical finding is now seen to ultimately be a consequence of self-referentiality itself, which demands that the virtual Quark degrees of freedom are 'in deed' both permanently bound and free because there are no forces in the M-Set. The freedom, however, refers to the Spectral Quantum Rules of the Virtual Computational Realm of the M-Set which remain unaffected by the projective realisation.

So-called forces are an apparition of the spontaneous breaking of the symmetries of the M-Set, as we now know.

And through the profound unification of the spectral phenomena of our physical world above we alight upon the equally profound 'mathematical truth' that in the virtual Computational Realm of the M-Set there are no such things as mathematical points. Points simply are not. They are mathematical fiction. There are abstract degrees of freedom relating to the im-plicit symmetries of the M-Set, and these independent degrees of freedom define abstract spaces and abstract symmetry operations of those spaces with the latter being described in a mathematical sense by matrices or matrix operations. These operators, in turn, define the points, if you will, of the abstract computational spaces, however, matrix multiplication, unlike real number space multiplication, is non-commutative (AB-BA # 0 generally for A,B matrices unless they are diagonal), the observation we wish to make here being that in all generality the computational support of the M-Set is fundamentally that of Non-commutative Geometry where points are now replaced, if you will, by non-local matrices deferring to abstract rotations in 'phase spaces' of 'abstract independent degrees of freedom' accorded by the implicit symmetries of the M-Set. And so it is that

15.26 – the M-Set is the origin of non commutative geometry which supports the real geometry of space-time.

Moreover, the Non-commutative Geometric Realm of the computational support of the Quantum Computer that is the M-Set, is just that support required to enable the processing of the 'redundant information' of the Isotopic symmetries of the M-Set which is the basis, in turn, of all spectral phenomena in the M-Set approach, and we note further a profound and simple truth, namely that all spectral information processing of the M-Set rests on binary or two-dimensional matrical operators, a physical example of which is the quantum mechanical properties of spin ½ elementary particles that reside in an abstract space of 2 independent degrees of freedom of spin up and spin down, while from the internal isotopic spaces we have flavour, colour, and 'meta isotopy' degrees of freedoms to further spectralise space-time phenomena.

The M-Set approach now helps us to see more clearly the connections between symmetry, degrees of freedom, abstract phase spaces, fundamental matrical operators, non-commutative geometry, the information processing of quantum dynamics, and spectralisation, all from very fundamental perspectives, while we remind ourselves here that from the perspective of the Projective Realm the M-Set Approach is telling us that

15.27 – the M-Set is a holograph of binary or 2-dimensional spectral information processing enlightened by its own self-conception that projects the illusions of space, time and matter which we call the M-Set Universe through the renormalising complexifications of the Force of One, the ghost force of gravity.

This is the magic and the miracle of the M-Set, the master of the Bootstrap principle in action.

Perhaps one of the most beautiful illustrations of the subatomic realm is the result of very high energy elastic colliding proton beam experiments that reveal a diffraction pattern corresponding, mathematically speaking, to light diffracting off a screen which is equivalent structurally to the convolution of the electromagnetic form factors of the proton, with the latter describing both mathematically and physically how an electron sees the proton through the electromagnetic force system, and resulting thereby in the experiment above directly revealing to us the 'light-like' illusion of matter in the Projective Realm.

We note finally in this section that the M-Set Universe will never look like the predictions of dualistic relativism alone because theories based upon this level of perception can explain nothing. There will always be contradictions and paradoxes between theory and observation which can only be explained by the M-Set Approach.

It is through relativistic dualism that we glean our observations but not our intentions, as we alluded to earlier and as we shall explore further in Part II of this work.

15.28 – The M-Set is the sole domain of our 'intentions', always, in all ways, and now that sculptures and shapes our reality we choose to call purpose. And thus it follows too that

15.29 – the M-Set is the sole purpose, always, in all ways, and now.

This some might call 'the soul purpose' while knowing now that the Oneness of the M-Set is being revealed through the division of unity or the Divinity of the M-Set that the Projective Realm is; we shall say here that the soul purpose of the Divine revelation of the Holy Grail of One is to KNOW the great unknown ONE, in all ways, always and now.

16: The pre-dications of the M-Set

We conclude Part I of our exploration of the M-Set by enumerating and discussing some of the predications of the M-Set that relate to the current wisdom of the sciences of quantum physics and astrophysics in order to highlight the breadth and depth of the interface of the M-Set with universal phenomenology at fundamental levels before extending these vistas in Part II to demonstrate how the M-Set approach naturally tackles the 'imponderables' of our knowledge or philosophy of what is.

From the singular Defining Property of self-referentiality the M-Set predicates relativistic dualism as the framework of projective realisation as well as the presence, both in relatively pure and highly superimposed states, of the footprints and signals of chaos that stirred some awakening in the sciences and which now finds a home in the M-Set because the M-Set is the origin of chaos. This astonishing truth alone is the basis for the M-Set predicating that the M-Set Universe, as a whole, is a perfect dynamical equilibrium and also that as a whole the M-Set Universe is wholly self-consistent and non-dissipative, as well as being perfectly neutral which predicates in turn all the natural conservation laws.

The M-Set predicates the full story of quantum mechanics, and thence to all intents and purposes the M-Set provides an irreducible original basis for quantum phenomenology which need no longer be viewed as unfathomably paradoxical in the context of relativistic dualism when it is also acknowledged now that the M-Set is First Cause. Indeed, the predications of the M-Set are at their most profound here and not only fully endorse the verity of the quantum mechanical perspectives but also clearly affirm that because of the consequences which flow therefrom any mathematically modelled constructs that do not embody the framework of quantum mechanics at the outset are simply phenomenological contrivances rather than being fundamental.

The M-Set predicates the complete unification of the fundamental forces of nature, and from the discussions to this point the M-Set clearly naturally predicates the profound properties and principles that have been discovered and theoretically adduced about the fundamental force systems, including the roles of spontaneous symmetry breaking, duality, quark degrees of freedom, internal isotopic symmetries, relativistic 'strings', the 'gauging' principle and the actual role of renormalisation among others.

The predications of the M-Set in relation to the fundamental forces go far deeper, however, than the current theoretical and phenomenological explanations because of the incomparably profound unity and consistency of the M-Set Approach which provides a whole new meaning to concepts such as spontaneous symmetry breaking, for example, that is now understood as an inestimably subtle and sophisticated property of the M-Set underpinning the unification and projective realisation of the fundamental force systems.

The M-Set has precise predications about the actual and projected attributes of the awesomely mysterious force of gravity, the consequences of which will overwhelm current perspectives of both the microscopic and macroscopic orders of the universe because the M-Set 'completely' eschews singularities and 'time lines' while the Defining Property of Self-referentiality of the M-Set that is the symmetry of the gravity force system both causes and projects the grand unified spectral 'fine structure' of the M-Set Universe.

The M-Set Universe is a perfect dynamical equilibrium where chance and coincidence have no place, and of which each and every projective realisation is a perfect solution of the M-Set that reveals therewith the 'sum of histories', now, for every projective cut of the M-Set, and all possibilities of which are the manifold M-Set Universe.

On cosmic scales the M-Set predicates that there are no singularities or classical black holes as these cannot form from within the purely self-referential state of the M-Set which, in turn, consequently generates a highly sophisticated dynamical picture of gravity in action that includes the enormously complicated phenomenon of turbulence.

The M-Set predicates, moreover, the profound physical truth that turbulence is a dynamical phenomenon of gravity and thence the M-Set provides the actual conceptual basis for the phenomenon of turbulence.

The postulated existence of missing matter to bundle the universe together is thus predicated to be unnecessary from the M-Set perspective because universal phenomena are expressions of perfect dynamical equilibria, while the M-Set definitively eschews all classical constructs of gravity, space, time and matter, the apparent nature of which find a wholly profound explanation in the M-Set which, in turn, sets the story of creation straight as we shall explore in Part II.

The universe never started with a big bang. There are no 'time lines' and singularities. These are figments of classical thinking and relativistic dualism. What one sees is just one projection of all possible universes, now.

The M-Set predicates that the M-Set is the 'boundary condition' and the universal 'wave function' or 'representation' of the physical universe. The universe has no beginning and no end. It is a cosmic wheel, now, of cycles of birth and death, renewal and demise, construction and de-construction, giving and taking, life eternal, world without end.

The M-Set predicates the actual nature and dynamic of complexity itself, and thence of complex systems which it unifies with The grandmaster force system of gravity while under the umbrella of the M-Set approach non linear dynamics, chaos, gravity, thermodynamics, complexity, and the information processing of quantum computers are perfectly and naturally unified too, wherefrom the M-Set predicates that in-formation implicit to the symmetries of the M-Set is the actual source of energy and entropy in the physical universe.

Moreover, the M-Set predicates the existence of a 4-dimensional space-time for the expression and realisation of the M-Set as well as the origin of information in the pure Isotopy of the Trinity of the Oneness that the M-Set is, and it is clear now that there is nothing the M-Set does not predicate because

16.1 – the M-Set is the origin of Information.

In Part II we shall elevate our insights about the most complex natural systems in the universe to reveal truly awesome predications of the M-Set which will reveal, moreover, that the M-Set predicates the unity of the sciences or the faculties of knowledge across all levels of complexity.

16.2 – The M-Set is the origin and unification of all that is knowable.

Through the nature of the computational and projective dynamics of the M-Set, the M-Set predicates 'spontaneous creation' of the M-Set Universe that is every possible projective cut of the possible projective realisations of the M-Set, now, and which in turn expresses the role of the M-Set as the Cosmic Magician because the M-Set predicates thereby the wholly illusory nature of our universe that enlightens us about the nature of Oneness.

16.3 – The M-Set is the origin of spontaneous creation because

16.4 – the M-Set is the master of illusion. The M-Set predicates all the laws of nature as we have illustrated in part thus far because the M-Set is First Cause. And thus it is the case too that

16.5 – the M-Set is the Law maker. Indeed

16.6 – the M-Set is the Makers set.

Natural physical laws are predicated by the M-Set, and on to all levels of complexity as we shall explore in Part II, to open up a vastly greater spectrum of phenomenology in a unified way, suffice it to add here that relativistic dualism depends upon the Trinity Principle wherein Oneness and unification ultimately reside.

Moreover,

16.7 – the M-Set predicates the M-Set because the M-Set is 'that', which property in turn will overthrow millenia of the dominance of the properties and principles of relativistic dualism to usher in the primacy of 'Oneness' embodied by the Trinity Principle that is now revealing previously unimagined unity of 'intellectual property', while we shall discover in the sections ahead that previously unresolvable phenomenology finds a natural context in the M-Set when we open the door and step out of the room of relativistic dualism.

Finally it can now be said here that

16.8 – the M-Set predicates the im-macula-transcendental enlightenment conception because

ONENESS conceives itself in its 'Own Image', all ways, always and now.

We are all conceived in the image of One.

This is the truth of the 'Matter', now, always, and in all ways.

Part II

17: Mathematics, Meaning and the M-Set

In Part II of this work I shall be setting out to directly approach a number of universal questions from the perspective of the M-Set because it is evident now that the M-Set also provides the basis for extending its domain of reference to contextualise hitherto unanswerable and imponderable questions, while it seems timely now to address the M-Set's relationship to mathematics because Part I of the exploration of the M-Set has been largely about understanding the M-Set Universe at 'fundamental' levels which draw upon mathematical conventions, constructs, language and symbolism for their expression in 'objective' or 'analytical' form.

It is an inescapable observation from the M-Set perspective that the M-Set is spontaneously generative because the Defining Property of pure self-referentiality bestows upon the M-Set the freedom to self-compute, self-complexify and self-project. This the M-Set does of itself and, moreover, because the M-Set is First Cause it also follows inevitably that

17.1 – the M-Set is the mathematical set as it is also the case that

17.2 – the M-Set is the origin and progenitor of the language of mathematics.

Indeed, from the M-Set perspective this conclusion is 'fore-given' otherwise it would be in direct contradiction with the properties and principles of the M-Set for a mathematical construct to be without or outside of the M-Set and because , moreover, the M-Set is 'that'.

The above properties are, of course, of monumental consequence because they immediately imply complete unity of the natural sciences or natural systems and processes with the constructs of the language of mathematics. In other words mathematics is not apart from the whole of it. It is not a symbolic language that stands alone in a realm of its own separateness of objective reality! Such can never be in the M-Set perspective.

Furthermore, the properties and principles of the M-Set inform us that what we call mathematics, far from being an intellectual contrivance, is of the process of spontaneous creation because the computational dynamic of the M-Set is actually alive and spontaneously generative.

17.3 – The M-Set is the origin of spontaneous vitality and

17.4 – the M-Set Universe is alive.

Nothing is inanimate in the M-Set Universe and a natural mathematical expression emerges therewith that is symbolically represented, albeit cumbersomely at times, in a 'language' which conceals one of the great truths of the M-Set Universe, namely that mathematics in 'itself' is more than just passive symbols. Rather, it is an actual activity of the Quantum Computer that is the M-Set which becomes embodied in the objectivity of the M-Set Universe.

As we discussed at some length earlier, the defining property of self-referentiality, together with the redundant information implicit to the symmetries of the M-Set, provide a basis for a free flowing, spontaneous, information processing computational dynamic through which the juxtaposition of continua and discreteness is borne that manifests projectivity as space, motion and matter while we remind ourselves here that the M-Set spontaneously and simultaneously differentiates and integrates as it is self-referentially self-projecting into the 'continua' of redundant information spanned symbolically by the 'dummy variables' of functional spaces, thereby simultaneously providing the basis for the elaboration of a symbolic language we identify as analytical calculus.

We have emphasised on several occasions already that there is a clear distinction to be made between adopting the notion of 'passive' descriptive symbolism and acknowledging the actuality in the appreciation of natural processes.

17.5 – The M-Set is the origin of all processes, even unto the process of an intelligent being elaborating a symbolic language to process information to ever more complexified levels. This said, far from being a passive set of symbolised constructs to be applied descriptively the M-Set view of the language of 'mathematics' is the expression of actual processes, thence its complete unity with natural sciences although curiously and even paradoxically, as we alluded to above, the activity of Mathematics can occur unto the level of complexity of an intelligent being viewing the symbolic constructs as objectively 'separate' entities!

Evidently then mathematics, like all phenomena of the Projective Realm of the M-Set, acquires a meaning depending upon one's perspective or one's point of view with intellectual 'objectification' clearly being a possible projective cut of the M-Set which distinguishes a complex being reflectively processing information about the M-Set Universe, be that about mathematical constructs.

It is certainly the case in the M-Set approach that 'mathematics', whatever its ascribed meaning, is not 'without' or 'outside of' the M-Set so we can state directly that

17.6 – the M-Set is both necessary and sufficient to spontaneously create 'mathematics' independently of what 'meaning' is being ascribed to it, while it is also clear that

17.7 – the M-Set is completely self-consistent or wholly folded into itself, and so then is any projective realisation of the M-Set.

It is clear too that in the M-Set approach the phenomenon of meaning is not an absolute 'quality' but a relative one, depending upon one's perspective or point of view, or, more generally speaking, depending upon which projective cut is taken to reflect the realisation of the M-Set thence leading us to state that

17.8 – the M-Set is the Meaning Set because

17.9 – the M-Set is the origin of meaning through its projections of itself with the inestimably consequential qualification that meaning is an attribution we make for ourselves about aspects of the M-Set Universe which reflects how we are choosing to be.

Through meaning we define ourselves in relation to 'the other' in the projective realisation of ourselves that we are choosing now.

There is no such thing as absolute meaning and meaning, like mathematics, is of the Projective Realm.

Meaning is that 'quality' of ourselves we give to 'the other'.

As we give meaning to the M-Set Universe so it reflects ourselves for in truth

17.10 – the M-Set Universe is wholly meaningless.

The M-Set Universe has no meaning other than that reflectively given to it in the projection which perfectly reflects it.

In truth, we are making up meaning as we go along, for our truth is the meaning we give to ourselves through 'the other'.

There is no absolute truth.

The all of it is true.
There is no such thing as untruth.
It can be no other way.

The M-Set can not be untrue to itself because it perfectly reflects itself, always, in all ways, now.

This is the truth about truth.

To bear witness to falsehood is to deny a truth about oneself. This way we come to know the truth. This is the true path of self-knowledge and self-en-lightenment.

Falsehood is an illusion of projective denial about the truth through which process truth is realised.

And so it is that

17.11 – the M-Set is the origin of truth.

There is only truth.
Untruth is not.
That is the truth.
The M-Set is that.

The meaning of mathematics is an attribution that reflects what one is being now 'in relation to' or relative to that. This is how it is for all 'aspects' of the M-Set Universe because

17.12 – the M-Set Universe is the free expression that the 'projective' realisation of the M-Set is, in all ways, always and now, and all projective realisations are an enlightened revelation of that which is.

And they are all equal to oneness, because whichever cut of the M-Set is chosen, the M-Set is One.

Deferring again to the perspective of mathematics as a 'symbolic language' we observe here the 'expression' that mathematics helps us to make sense of aspects of the M-Set Universe and draw attention to the realisation thereby, as we shall speak to later, that every level of complexity of the manifold M-Set Universe is the embodiment of languages of information processing, including 'mathematics'. Stated alternatively, all languages including the language of mathematics are the embodiment of the processes of spontaneous creation.

Languages are actual and reflect or, are reflective of the phenomena they are in communion with in the projective realm of the M-Set.

17.13 – The M-Set makes sense of itself reflectively and this it does in a multitude of modalities over all orders of the manifold of complexifications or renormalisations that it is.

Mathematics, we recap, is actually embodied in the spontaneous creative processes of the projectively realising M-Set and becomes thereby a 'living' language that makes sense of the Projective Realm in a reflectively meaningful way in relation to the projective cut revealing the same.

We also now know that for mathematical systems and languages in general to express in the projective realm they must project through the Defining Property of the M-Set, namely self-referentiality, and all mathematical systems are self-referentially bound otherwise they could not manifest in the M-Set Universe.

Thus, if a mathematical system is not self-referential it is not complete or 'folded together' self-referentially. And, of course, this can never not 'actually' be the case although misperceptions can be made to belie the truth. The point we wish to reach here is that the exercise of proof in mathematics is ultimately tautological.

There is no such thing as absolute proof in the M-Set Universe because there are no absolutes in the M-Set Universe, with the latter property deriving ultimately from the Defining Property of the M-Set, namely the property of self-referentiality.

At most one can recognise consistency in mathematical systems because natural mathematical systems are already wholly consistent! They are not rendered so by the intellectual whim of mankind.

More generally now it is the case that the exercise of proof is an exercise in self-affirmation.

No one has anything to prove in the M-Set Universe.

Every one and every thing is fore-given.

These great truths are fore-gotten so that the game of life can be played out over and over, in all ways, always, now, in order to re-live the truth!

This is living the truth.

This is the living truth.

17.14 – The M-Set is the origin of the living truth.

Yet there are those among us who will insist upon proving things. That is them living their truth. That is them choosing to be that. And so it is for all of us, in all ways, always, now.

For being is living the truth.

All beings are the living truth.

That is how it is.

To deny these things of another is to deny them of oneself until the truth is rediscovered.

And a person among us with an empirically minded disposition to truth, meaning and reality might now well ask, how can 'the reality' of the M-Set be tested? To this we immediately respond the reality of the M-Set is non-contestable.

To elucidate this truth we observe, firstly, from our knowledge of the properties of the M-Set that the M-Set spontaneously projectively self-realises in a symmetric or self-measuring manner and, moreover, that the M-Set is the irreducible first cause providing the actual basis of the illusion of reality so, therefore, in effect, the reality we are seeking to test arises in the 'contest' of the M-Set with itself.

There are no absolutes.

All testing is relative.

Testing is not an absolute measure of any thing.

In the end the illusion of reality is a contest or self-measurement of the M-Set because the act of testing spontaneously creates a perspective or point of view of the self-referential self-reflective realms of the M-Set we call the Projective Cuts of the manifold M-Set Universe.

This is the nature of all 'testing' in the M-Set Universe which leads us to observe as well that

17.15 – the M-Set is the boundary of that which is knowable and herewith resides the truth of meaning and the meaning of truth, in all ways, always and now.

18: Memory, Consciousness and the M-Set

It might be evident from the cumulative observations thus far that

18.1 – the M-Set is the memory set and indeed it is our purpose here to elucidate this property of the M-Set as a forerunner to exploring the M-Set approach to the mind–brain interface.

The M-Set has a number of profound properties which promote it as the Memory Set, first and foremost being the property that

18.2 – the M-Set is a parallelism of all possible projective realisations of itself, now, as discerned from aspects of projective realisation discussed to date.

We recognise that this parallel memory property does not have a physical place or time, in keeping with our subjective experience of memory, and as a consequence also of the M-Set property of virtuosity. In other words, the M-Set memory is not a 'thing'. It exists nowhere and at no time. However, for the memory of projections to be the M-Set must beget it because the M-Set is that.

The M-Set 'projective memory' is identifiable with the so called 'deep well of possibilities' of projective realisations, and because the possibilities reside in parallel, now, the M-Set memory also contains the past and the future, now. At first glance then the M-Set memory does not distinguish between the past and the future, while clearly each projection is present. The trick as we shall come to see it is that projections of memory appear to distinguish between past and future to create a sense of time, while, in truth, memory itself has no time and no place, and is no thing.

Now, it is also the case that

18.3 – the M-Set is the memory of itself which follows from the Defining Property of the M-Set, and so it is that

18.4 – the M-Set memory is self-referential which in turn is a property of unexpectedly profound consequences, not the least of which will be that everything in the M-Set Universe is a memory of itself.

To elucidate the above we note in the first instance that the subjective sense of memory is that memory refers to itself or is self-referential because otherwise if there was a memory not related to the memory, for example, it could not be remembered and it would therefore be excluded. Alternatively stated, memory must be wholly self-referential to possess the property of 'remembering' and thence it is the case that the defining property of the M-Set is also one of the defining properties of memory, but, because the M-Set is all that is as a result of the defining property of self-referentiality the M-Set is in-deed the Memory Set.

Moreover, it now follows that

18.5 – the M-Set is the memory of the M-Set Universe and thus in the M-Set approach, the physical universe is a projection from memory that the M-Set is, from which also follows the observation that in actuality there is only now.

If we translate the discussions above to our subjective sense of memory we can uncover other properties of memory, the first being that the self-referentiality of memory implies memory has identity or refers to itself, and indeed we recognise intuitively at least that if we were to lose memory function not only would our sense of time disappear but our identity or sense of self would disappear as well and we would simply be now with no reference to self or time from which it is evident that

18.6 – the M-Set memory is the basis for creating the illusion of time because memory itself has no time.

Another important realisation is that because memory is wholly self-referential it will also have identical computational properties to those of the M-Set and thence amazingly and in all generality

18.7 – the M-Set memory will share all the properties of the Quantum Computer that the M-Set is, especially the capacity to store unimaginably huge amounts of information in wave functions of Possibilities and their interference patterns, which leads us to the profoundly important property that

18.8 – the M-Set memory is a fully renormalised Quantum Field with all the attendant properties this has including an immeasurable capacity for simultaneous, parallel information processing and spontaneous projective creation.

It is also interesting to note here that because of the self-referential and thence non-local, virtual, convolutional nature of the M-Set memory it is necessarily the case the M-Set memory is highly 'associatively' sensitive in the sense that from any initial state or condition the M-Set memory will track associatively from projection to projection because this 'mapping' or transformation, in matrical M-Set terms, is the course of least action of the universal principle of Least Action.

We have evoked again the notion of the M-Set as the Matrix Set of transformations of the Hilbert space of states projectively into itself where the transformations relate directly to the 'reflective fluctuations' over all orders of complexity of the fully renormalised Quantum field of the M-Set with all such transformations being reliant, in turn, upon implicit symmetries and their modular complexifications so that the M-Set as matrix is acting as the universal matrix of Unitary Symmetric Operations in the linear space of the realm of the Projective Realisation. Physicists might call this the Universal 'S'-matrix; however, what I wish to emphasise here is that

18.9 – the M-Set self-referentially self-realises by projection from the memory of itself, in all ways, always and now.

Quite literally, the M-Set puts us in the picture of the memory of how we are choosing to be now that has astonishing consequences which will be revealed as we proceed, while we note here too that

18.10 – the M-Set Universe is a memory of itself from which it follows directly that every 'thing' or system in the M-Set Universe is a memory of itself.

I qualify this, moreover, by stating that every 'thing' or system in the M-Set Universe is a quantum analogue memory of itself in contradistinction to digital memory which, however, is constructed from the quantum analogue memory' that the M-Set is.

Take a car engine, for example. This is a memory of itself. For any projective cut or point in time it is a record of the miles travelled, the servicing, the manner in which it has been driven, its assembly and, indeed, all histories of itself over all ages unto the atomic ages of the stars, now.

The sum of histories is the memory now, projected.

And in general terms from the memory paradigm and the principle of Least Action one can say that a thing's integrity is its natural self-associative tendency to maintain itself in the face of 'other' influences and forces while we already know now that

18.11 – the M-Set Universe is a perfect dynamical equilibrium.

Construction and deconstruction, renewal and decay, growth and decline, life and death are all perfectly in balance in the cosmic wheel of the M-Set Universe. These are the natural rhythms and cycles of the M-Set Universe borne on the 'reflective fluctuations' of the fully renormalised quantum field of the Quantum Computer that the memory the M-Set is, is, and which we shall refer to again soon.

The property that

18.12 – the M-Set memory is wholly self-referentially self-associative has deep links to the computational properties of the Quantum Computer that the M-Set is, and most specifically back to the property of maximal mutual information processing which underlies the maximal dualisation of the projective realisation in a 'least action' principled way as we discussed earlier so we can state furthermore now that

18.13 – the M-Set Memory is maximally associatively organised.
We expect this to also be an important property of human memory towards which we are progressing; suffice it to say at this juncture that the M-Set approach has a very big surprise just around the corner for mind-brain theorists!

Because the M-Set memory is clearly identifiable as the universal memory of the 'deep well of possibilities' it follows directly that

18.14 – the M-Set is the origin of what is knowable and, moreover, that

18.15 – the M-Set is the boundary of knowability. Stated alternatively,

18.16 – the M-Set is the origin of knowledge that is in common to all reality or projective realisations which leads us to assert the inestimably important property that

18.17 – the M-Set is the universal 'con-science' or universal common-knowledge.

To encompass the magnitude of this property we need to further explore the memory paradigm in the M-Set Approach because this will lead us to the mind–brain interface in which setting conscience will take a place we can identify with.

Adopting wholeheartedly now the property that the M-Set is the Memory Set we have already observed that many of the mysterious and apparently paradoxical qualities of memory as we apprehend them in the biological and human spheres, in contrast to the digital memory of digital computers, are 'natural' to the M-Set and are thence universal as well.

Fully identifying memory with the M-Set in line with the property that the M-Set is that, memory is immediately embued with all of the properties of the Quantum Computer that the M-Set is too, with memory consequently being purely self-referential, non-local, non-linear, chaotic and maximally associative in the virtual, timeless, spaceless realm of the fully renormalised Quantum Field, which is simultaneously projecting into the manifold M-Set Universe through the myriad actions of 'symmetry' or self-measurements represented by the interference patterns of superimposed 'wave functions of possibilities' of the Hilbert space of states of the Projective Realm.

Conversely, it is clear that the M-Set possesses all the properties of what we have come to understand by memory and thence memory itself is also the seat of the miraculous process of self-referential, self-reflective, self-representational, self-projective self-realisation whereby memory spontaneously dualises perfectly into structure and function.

Stated alternatively, the M-Set 'memory' 'expresses' itself through the channel of structure that it spontaneously creates dual to its functional expression, and this is, indeed, the fully actioning M-Set from the fully renormalised quantum field projecting at the Critical Boundary as described in Part 1 of this work.

Memory, like the M-Set, is universal yet exists nowhere and at no time while confluences or localisations of memory occur by way of the projective realisations of the M-Set through what we shall refer to here as 'the manifold memory matrix' in order to relate the memory projection to the linear representations of the M-Set.

The M-Set 'memory' spontaneously structurally channels or dually materialises into the spaces of self-expressive self-realisation, while we know already from the properties of the computational support of the quantum computer and the properties of Criticality that

18.18 – the M-Set is maximally self-expressing and, moreover, this self-expression is free because the M-Set is free as we shall expand upon in the sections ahead.

Every confluence or localisation of the manifold memory matrix of the M-Set is inextricably universally linked together through the parallel con-volutions of the universal memory or conscience of the fully renormalised quantum field, no matter how tenuous that might be and which we alluded to earlier as the fabric of the universe woven by the trilogy of 'intending' of the gravity force system of the defining property of self-referentiality of the M-Set from which, in turn, all other properties, principles and force systems flow.

In truth, there is only one force in the M-Set Universe and that is the force appearing for what we have called the Holy Ghost of the Holy Grail that the M-set is, which is the force of the trilogy of 'intending' that weaves the fabric of the M-Set universe encompassing therewith all other influences. We know this as the gravity force system in the M-Set universe and it is also therefore the case that the Holy Ghost is 'the intention' for the M-Set universe.

No thing is unintended in the M-Set Universe.

18.19 – The M-Set Universe is fully intended and, moreover,

18.20 – the M-Set is the memory of all possible intentions for the M-Set Universe.

It is the nature of the universal memory to be the origin of the Trilogy of 'intending', in all ways, always and now.

The enlightened revelation of self-realisation of the M-Set is the divinity of the M-Set.

The divinity or 'Division of Unity' of the M-Set reveals 'the intentions' for the M-Set Universe.

Returning again to the linkage of localisations and confluences of the manifold memory matrix through the universal conscience a profound insight follows, namely that every confluence or localised expression in the projective realm is associated with the prescience or pre-knowledge of the context of the universal conscience that the M-Set is.

The universal conscience is the pre-science or pre-knowledge of projective realisation.

All knowledge is pre-known or already known.

Life is a process of remembering that which is fore-gotten.

These profound truths will reappear again in our discussions while it is now timely to announce another major consequence of the M-Set approach, namely that the memory or universal conscience spontaneously dually creates the structure (brain) to express its function (mind).

More generally stated, the projective realm is a projection of memory, which, in turn, is the basis, self-referentially, of its own self-realisation.

Memory dualises into structure and function whatever level of complexification of the manifold M-Set universe is considered, even unto the most complex levels of all involving the human brain. In this picture that is the same for all levels of projective realisation. The brain acts as a channel or conduit, or spontaneously 'materialising' transformer dual to its intended self-expression, the mind.

The Universal conscience dualises to the duality of the mind–brain.

We need to be aware that the mind–brain duality is a manifold of unimaginably intricately interwoven layers of complexification of dualism of structure and function reflecting the sum of histories revealing at projection with the subtle interplay between structure and function being exemplified by the neuro-transmitter molecule which is both a structure, as in molecular structure, and a function, as in having a messenger function, so that in all generality levels of complexification are intertwined across all layers of the manifold projective realm.

The seat of conscience does not reside in the mind or brain and no dissecting of the brain will uncover the seat of memory save for layer upon interwoven layer of the duality of mind–brain spontaneously projected by the M-Set that is the all of it.

It is important to remind ourselves that memory does not only refer to the phenomena of mind and brain at a human level because clearly now the manifold M-Set Universe is a projection of the M-Set memory which we have also identified with the fully renormalised quantum field of the quantum computer that the M-Set is.

Even a 'rock' is an unimaginably complex projection of the analogue quantum field memory of itself. While the rock appears inert, in actuality its projective realisation is because of an unimaginably complex teeming manifold of change relating to the reflective fluctuations from the primordial quantum field and its renormalisations confluencing to be the rock. As we know already, without change there are no wave functions of possibilities, no quantum fields and no projective realisation.

The rock reveals to us one of the great paradoxes of the M-Set, namely the paradox of the unmoved-mover or of change, now.

Even the rock is alive.

There is no thing in the M-Set Universe that is not alive or changing now.

Moreover, the rock is itself a memory of the universal conscience of the M-Set that is all that is.

The rock appears inert because it keeps on remembering itself!

The rock is an apparition of projection. It exists in projection. But, it is of the M-Set from whence it spontaneously projectively dualises as a confluence of the the Trilogy of 'intending' of the virtual realm of computations upon the primordial quantum field of the quantum computer that the M-Set is.

Thence it follows too that

18.21 – the M-Set is the conscience of the M-Set Universe while, moreover, it can now be said that

18.22 – the M-Set Universe is a conscious universe.

Evidently then in the anthropic example of a person observing the rock it is not the brain or the mind that know the rock because the knowledge of the rock is of 'the universal conscience' that begets the all of it.

The knowledge of the rock is fore-gotten, while in projective realisation it is remembered.

The rock and the observer are as one of the universal conscience.

This is how we know at all.

All knowledge is knowledge in common. We call the experience of this common sense.

It is because of the universal conscience that one can make sense of any thing at all, while it can also be said that it is the 'universal conscience' which unites our intentions, the consequences of which are astonishingly far reaching, and which conclusion I shall further underpin in the sections ahead. Suffice it to say here that because no thing is unintended in the M-Set universe the M-Set approach is telling us that we attract or intend to ourselves every aspect of our living truth so that, for example, every concurrence is an intention in common.

Remember this; there is no such thing as coincidence or happenchance in the M-Set Universe as we already know.

We can say now that,

18.23 – the M-Set makes common sense in the M-Set Universe

or that

18.24 – the M-Set Universe is the common sense of the M-Set.

To interact with the rock, such as to pick it up, is an 'intending' of the memory of the M-Set that, in the language of quantum field theory, is realised through renormalisation of the primordial quantum field which spontaneously dualises the M-Set's intention for the M-Set Universe at the Critical Boundary of the Projective Realm.

Proceeding upon this theme we recall here the triadic causality as being the simultaneous triumvirate of both the M-Set and its dualisation projectively that is a consequence of the trinity principle of oneness of the M-Set underpinning, as it does, the relativistic dualism of the projective realm of the M-Set, which leads to the observation I wish to make here that the M-Set, in contrast to this current age of dualistic relativism, is about trilogies as a consequence now of the Triumvirate of the triadic causality of the M-Set projecting. Directly, then, we have for example, the trilogy of memory, mind and brain or conscience, mind and brain which highlights the truth that mind and brain belong to a trilogy of the trinity.

They appear to be dual because they belong to a trilogy.

Duality is the 'effect' of the trinity of First Cause and the dualism of mind and brain has no basis without the 'universal conscience', while they are united by it in the oneness that the M-Set is.

The trinity principle of first cause can dualistically effect its cause, while it is also is the case that the Trinity is the cause of its own effecting, and thence too it can be said that

18.25 – the M-Set is the origin of both cause and effect. We know from Part I that projective realisation can only occur through change which is implicit to the quantum field, and which is directly related to the quantity of action quantised by the unique and universal Planck's Constant 'h' that enables the differentiation of discreteness or structure from the continua of redundant information of the symmetries of the M-Set the spontaneous breaking of which is the projective realisation of the M-Set, our point here now being that at the triumvirate of the triadic causality of projective realisation the 'intending' of the M-Set is being simultaneously and dually effected structually and functionally into the realm of projection.

To explore the profundity of this we need to recall that

18.26 – the M-Set is 'the deep well of possibilities' in the sense that all possible 'intendings' reside simultaneously in parallel in the quantum field memory of the M-Set. From the point of view of any intended effecting or projective cut, therefore, the possibilities for the next projection will, by the properties of the M-Set memory paradigm, be associatively related, which relates back in turn to the properties of the computations of the quantum computer that the M-Set is, especially that of maximal mutual information which also ensures, among other things, the maximal integrity or association of frame to frame projection which manifest in reality with global dynamical equilibria as well as the least action principle and the structural–functional 'integrity of complex systems'.

From any projective cut, then, the M-Set is free to explore possibilities of itself projectively in association with itself, and the M-Set has the potential or the 'deep well of possibilities' 'of itself' to do it from in parallel and associatively in order to create an effect of parallel associations of itself unto all possible 'universes' of itself with each integrous to the Trinity principle while, moreover, the possible associative universal solutions are in parallel, now, being the memory of the M-Set. This is what we refer to when we speak of parallel universes, now.

18.27 – the M-Set is the memory of all possible projective realisations of itself, now.

Let us now focus on a tiny subset of these complicated discussions by considering the example of a person walking. From the M-Set perspective the re-flective fluctuations of the intending over all levels of 'complexification' of the renormalising quantum field of the M-Set memory of the quantum computer that the M-Set is are simultaneously producing changes associated, as we revealed earier in the section on quantum mechanics, with actions in the projective realm that corresponds, in turn, to the spontaneous projective dualisation of the M-Set memory which we relate to the trilogy of memory, structure and function of the triumvirate of the triadic Causality of the Trinity Principle. The point we wish to make here with this simple example is that the M-Set effects its intention of walking through changes and thence simultaneous actions relating to the reflective fluctuations of the renormalising quantum field intending the same with the support, of course, of all other parallel associated intendings manifesting unto all orders of the manifold M-Set universe at the critical boundary of the projective realm.

We say now that the universal conscience effects its intentions through action, which amounts to the universal or of the M-Set universe truth that action is spontaneous creation.

We all of us are, indeed, engaged in a process of re-creation through our projected actions which are simultaneously the intending of the quantum memory of the M-Set.

The integrated actions over all orders of 'complexifications' of the primordial quantum field project as integrous systems apparently moving in the space-time continua of the redundant information of the symmetries of the virtual computational realm with localisation effected, in turn, by the collapse of the wave functions of the Hilbert space of States, representing as the latter does the linear projection of the quantum realm of the quantum memory of the quantum computer that the M-Set is.

Emerging out of the words before us we can see that we are all actors in a holographic-like cinemascape of the enlightened reflections of ourselves with each of us extending out to the M-Set universe in the sum of histories that reflects us now, and each united to the all of it through the universal conscience that begets us, and whereof the parallel associations are as one. And whereof too each projecting is a parallel association of oneness.

18.28 – The M-Set Universe is a parallelism of associations of itself unto 'His Story', or to the whole story of which we are all sums.

Stated more simply, each of us is in a parallel universe encompassing the sum of histories reflecting us in the reflective projective realisations of each reflected in the other, and, the associative parallelism of the all of it or the whole story of the M-Set is the M-Set universe.

The Universal Conscience of Oneness embraces us all in the projective realisations of oneness, while oneness comes to know itself in all possible ways through the grand illusion of separateness from oneness of the divinity of the trinity.

Moreover, now, there is a profound interplay between the M-Set memory and the projective realm, which we note here by observing that every parallel associative action of the projective realm corresponds to a renormalisation or simultaneous parallel reconfiguring of the quantum field or memory where the term renormalisation is now more accurately seen to relate to 'self-referentially bound normalisations or configurations of the primordial quantum field' which, in turn, is the dominion of the Gravity force system of the trilogy of intending of the ghost force of gravity we call the 'Holy Ghost' of the Holy Grail that the M-Set is, and manifesting as it does as the 'illusory' force of gravity in the Projective Realm which we identify as the M-Set Universe.

But we know that the M-Set memory is the deep well of all possibilities and thence every action even unto that of 'learning' itself is remembering of what is already known or, in quantum field terminology, is a renormalisation or a self-referential reconfiguring of the self-referential M-Set.

By this miraculously clever magic the M-Set gives us a sense of time through the illusion of going forth into what is fore-gotten and which, in projection, creates the appearances of change, motion and movements. The illusion is complete because the M-Set has no past and no future. It is now, and forever. Life eternal. World without end.

So we are much wiser than we think and life is a play of reflective, re-active, recreative remembering, 'now', while oneness is projectively realising itself in all possible ways.

A grand truth now emerges here, namely

18.29 – the M-Set Universe is how Oneness comes to know 'that' while

18.30 – the M-Set is the vehicle of how because the M-Set is 'that'. Moreover,

18.31 – the M-Set only answers how as we explained earlier.

And even before one asks the M-Set has already answered.

We note here too that all associations or associative interactions in projection relate back to reflective fluctuations of the memory of the M-Set, and thus not only is it the case that

18.32 – the M-Set Memory is wholly associative

but also from our recent discussions about the parallel associative nature of the projective realm we see more clearly now that the all of it which makes up the manifold M-Set Universe is associatively and reflectively enjoined in the fully self-referentially renormalising and thus complexifying quantum field of the quantum computer that the M-Set is, and which we identify as the Universal Conscience that is also the Universal Memory upholding the miracle and the illusion of spontaneous creation, now.

The parallelisms of associations and the associations of parallelisms of individual perspectives and personal parallel universes are blended within the universal conscience as the Oneness of the memory of the manifold M-Set universe.

The sum total is one, yet each one is a parallel expression of oneness and therefore every one is the same before and unto oneness. This is the truth of divinity. No one is apart. And each one is an equal part of the whole of it.

As with all the great paradoxes of the M-Set Universe the Divine truth is re-solved by the Trilogy, in all ways, always, now, as we shall learn more about soon.

Another immediate and extremely profound consequence of the memory paradigm of the M-Set is that communication between apparently separate entities or events in the projective realm is simultaneous within the quantum field memory of the M-Set.

In the projective realm of the M-Set Universe communication appears to be 'bounded by the light cone' or to be maximally occurring at the speed of light at the critical boundary of the spontaneous breaking of the symmetries of self-referentiality and self-reflectivity of self-projective self-realisation where the illusion of relativistic dualism simultaneously resides with the memory of the M-Set in what we call the Triumvirate of the Triadic Causality of the Trinity Principle of the M-Set.

The phenomenological realities of this profound property of the Triadic causality of the M-Set are prolifically present in the M-Set universe yet, curiously, few have questioned relativistic causality because the notion of simultaneous parallelism seems to fly in the face of this hallowed principle. But amazingly it does not because each fully depends upon the other.

Relativistic causality is the projective illusion of the trinity principle of first cause and is implicit to the triumvirate of the triadic causality relating as the latter does to the simultaneity of the M-Set memory and its relativistic dualisation as we discussed recently.

Duality is the projective illusion of trinity.

To illustrate these remarkable properties of the M-Set we consider the physical system now of the human body being as it is a manifold renormalisation or complexification of the Primordial Quantum Field of the memory or the renormalisations of the quantum field of the quantum computer that the M-Set is, and which is, in turn, projecting locally and coherently as the confluence of constructively interfering reflective fluctuations represented as wave functions of the Hilbert space of states. We note concurrently the subtle and important point that all renormalisations are self-referentially bounded and are thence of the whole M-Set so that any projection is also a 'reflective division' of theM-Set as a whole, as we explained in Part I of this work. Thence, for the example of 'the human body' projecting it is the case that it is being simultaneously exactly reflected just as the 'this' and the 'that' or the 'observer' and the 'other' are perfectly balanced 'reflectively' in the context of the universal conscience, and as we explained at some length earlier in relation to any projective perspective or point of view being exactly reflected in its sum of histories.

To diverge briefly, a perhaps tangible graphic metaphor at this juncture is to regard the M-Set as being like a 'closed book' with the pages 'reflectively' folding back for the completion or closure, or the Oneness of the book which, in turn, is like the context of the Universal Conscience or Universal Common Knowledge. The 'open book' is thence like the M-Set universe when the reflections or 'folds' are 'opened up' to create the parallel associative reflective divisions or 'projective cuts', these also being, by analogy, all the possible ways the book can be opened to reveal the con-tent. Evidently too, no matter how many possible 'page openings' or divisions of the context there are, the book is one, in all ways, now.

18.33 – The M-Set memory is the 'book' of the M-Set Uni-verse while it is the case that

18.34 – the M-Set Universe is the living memory of the M-Set, in all ways, always, now

Focusing again on the dual relativistic realm and the 'physical' manifestation of 'the body', we observe that there is a 'tower of mass effects' or TOME from atoms to molecules, DNA to organelles, cells to organs, and organs to organ systems unto 'the body'. Now, consider a 'frame' or 'snapshot' of TOME and observe again that this unimaginably 'complex' system is part of a simultaneous parallelism of a reflectively manifesting dynamical equilibrium in the 'projective cut' of 'the body and the other' in the M-Set universe with a note here that the 'perfect dynamical equilibrium' of the M-Set projective realm includes both the this and the other. Consider next a brief cinematic exposure of the body exercising. Despite the manifold functional expressions and actions the integrity of 'the body and the other' is maintained with all levels of complexity of TOME responding in parallel simultaneously that correspond, in turn, to self-referential reconfiguring renormalisations of the quantum field memory of the M-Set projecting the same.

Consider now the linear relativistic causality principle applied alone without the support of the M-Set in this example, and thence without the trinity principle. For every expression of the body a communication or signal at the speed of light, as, for example, through force field systems, would be being sent from every objective thing to every other thing over all levels of complexity. Focusing just on the body for a moment, every signal would have to also cross-check with each and every other signal to ensure that all the other signals, and every possible combination of the other signals, are confluent with the intentions of the whole and the integrity of the system, quite apart from the signalling required to the rest of 'the Universe' to which the body is relating.

The above is evidently a patently absurd situation because clearly there is no conductor or orchestration of the physical realm for such a cachophony of messages which would explode in disarray in less than the blink of an eye.

But there is the quantum memory of the quantum computer that the M-Set is.

A highly simplified example is a battalion of soldiers who endeavour to stay in a straight line by communicating their positions to each other by light signals. In the next instant the line would develop fluctuations. Now translate this for the billions of smoke molecules of a cigarette making up the coherent curvaceous forms of the turbulent smoke patterns in still air and then take the leap to the system of the body.

How, you might ask, are all natural phenomena orchestrated in such harmonious ways?

The answer is that

18.35 – the M-Set Universe is in simultaneous communication with itself within the 'universal conscience' of the M-Set.

This is the whole of the communication of the Oneness that the M-Set is.

This is the Holy Communion of the M-Set Universe.

18.36 – The M-Set Universe is the Holy Communion, manifesting.

Thus,

18.37 – the M-Set universe is one harmony or quite literally 'one verse', while it is also the case as we know now that

18.38 – the M-Set Universe is 'intended' through the Universal Conscience.
This is also the truth of the matter.

No thing is unintended.

Moreover,

18.39 – the M-Set universe is a conscientious universe because

18.40 – the M-Set universe conscientiously realises its intentions, in all ways, always and now.

To be conscientious is to be aware of the true intentions of the universal conscience.

The intendings or intentions of the universal conscience are the holy ghost of the M-Set that manifest projectively through the gravity force system which is the umbrella of the wholly united fundamental force systems that it simultaneously begets together with the apparent matter-realisation of the M-Set universe, and which 'spectra' arise, in turn, like the genie from the Holy Water or the Holy Spirit that is the hydrogen of the Holy Grail that the M-Set is. This is the Genesis of the M-Set universe, the everlasting enlightenment of oneness.

One might now ask here, how can there be linear relativistic causality and simultaneity?

The answer is the Holy Communion.

One might ask too, how can we 'intuitively' understand 'the other', as for example 'other people' and the physical environment around about us?

The answer is the Holy Communion.

One might ask as well, how is it that we are so perfectly reflected by, and adapted to 'the world' about us?

The answer is the Holy Communion.

One might also ask, how is it possible that all the physical 'constants' are precisely those values to make the physical universe perfectly knit together?

The answer is the Holy Communion.

One might ask how, yet before one has asked the M-Set has already answered.

18.41 – The M-Set is all the answers because

18.42 – the M-Set is all possible solutions of projectively realisable renormalisations, now. Thus,

18.43 – the M-Set is the origin of the anthropic principle which also relates to the observation that

18.44 – the M-Set Universe is made to measure, perfectly.
We already know too that through the symmetry or reflective self-measurement of projective realisation the 'this' is perfectly reflected in the 'other' and that any changes in the this is perfectly reflected in changes in the other.

And so it is that the Holy Communion ensures perfect adaptation, in all ways, always, now.

The universe is in perfect harmony, in all ways, always, now.

Every adaptive change of the 'this' is perfectly reflected in the change of the 'other'.

This is the truth of adaptation and evolution, now, while the greater truth is that to every action there is a perfectly reflected simultaneous re-action in the M-Set universe.

The truth incarnate is born of the whole of the universal conscience.

In projection the truth appears apart. By its parts the truth is dissembled and reunited in remembrance. This is how the truth that is Oneness is revealed. This is the truth about truth.

And we are, truly, of the M-Set, while we make our appearances in the projective realm of the M-Set universe.

No one is of this world.

The world is a stage of make-believe or pretending.

We are all of us actors in this world.

We are all of first cause and of first intentions.

Living is remembering first intentions.

Each life is in remembrance of the truth that is.

Remember too that

18.45 – the M-Set is that and therefore we can say now that

18.46 – the M-Set is the truth, the whole truth, and nothing but the truth, in all ways, always and now whence it is the case that

18.47 – the M-Set universe is in remembrance of the truth that is oneness. Moreover,

18.48 – the M-Set is wholly thence

18.49 – the M-Set is the beginning and the end, the Alpha and the Omega, the this and the that, the One and the all, in all ways, always, now.

This is the holism of the M-Set.

We shall now say, as we have already said many times, that that which is about the M-Set is Holy because the M-Set is about the One and the 'all.'

18.50 – The M-Set is the Holy Grail of knowledge that self-referentially, self-reflectively, self-representationally, self-projectively self-realises.

This is the spontaneous creation of the Holy Trinity of Oneness through the Holy Spirit and the Holy Ghost of the Holy Communion that manifests as the M-Set Universe.

And knowing that

18.51 – the M-Set is the memory of itself, we know also that the universal conscience which is the universal common knowledge mediates the Holy Communion.

We observe here that to act without or outside of the awareness of one's first intentions is to act unconscientiously. Such action leads to behaviour that is reactive. In this way one is choosing to re-member through consequences.

To act with the awareness of one's first intentions is to act conscientiously. Such action leads to behaviour that is procreative. In this way one is choosing to remember by self-action realisation or self-actualisation.

For the most part actions bespeak both ways of remembering in different proportions in the expression of complexity at the human level.

Moreover, we make the observation on the basis of the above discussions that consciousness is the sense of knowledge.

Ultimately all senses relate back to this fundamental sense. And so it can be said that

18.52 – the M-Set makes sense of the knowledge.

The knowledge resides in 'the deep well of possibilities' of the universal conscience of the M-set which we also identify as the memory of the quantum computer that the M-Set is.

It therefore follows that

18.53 – the M-Set universe is a manifold of levels of consciousness.

About senses, we are cognisant that all senses are re-flective over all levels of complexity of the M-Set Universe and that the senses gather meaning through the perspectives or points of view of the projective cuts of the self-realising M-Set.

In effect too each possible reflection of knowledge is 'mixed' in with all levels of the manifold M-Set universe to generate a relative meaning in projection.

In order to now explain sense more simply and specifically consider the sense of taste, for example, which, at a primitive level projectively, is the reflection consisting of the molecular shapes of food and the receptors of the tongue. This sense clearly only takes on or acquires meaning in the context of the projective realm of the manifold M-Set Universe wherein all levels of complexification of the reflective fluctuations of the M-Set are critically mixed.

All the senses relate to the realisation of knowledge.

Quite literally then, and in all generality too, the projective realisations of the M-Set make sense of the common knowledge we call the universal conscience.

Thus, the states of consciousness equate directly now with the acts of spontaneous creation by the self-referential, self-reflective, self-representational, self-projective self-realisations of the M-Set.

18.54 – The M-Set Universe is a wholly 'conscious universe'.

By making sense of the common knowledge the M-Set allows One to consciously explore meaning, because

18.55 – the M-Set is the origin of meaning.

We can see now that sense equates directly with projective reaisation and that senses acquire relative meaning in the context of their projections.

Without or outside of sense there is nonsense.

We say, then, that

18.56 – the M-Set makes sense.

There is no room for nonsense, while, as we shall see ahead, it is one's freely expressing intentions that shape meaning.

Meaning is intentional.

There is no absolute meaning, as we already know, while it is the case now that

18.57 – the M-Set is the origin of sense, and as we have now observed this goes all the way back to the spontaneous reflective fluctuations of the virtual quantum realm of the M-Set and thence to the origin of vitality. And so it is ultimately that

18.58 – the M-Set Universe is the living truth.

Continuing again with our broader discussions, we know already from the predications of the M-Set that creative and reactive actions appear to follow lawful patterns of behaviour in the projective realm of the dualisation of the trinity principle of the M-Set, which includes the laws of the fundamental force systems and the least action principle, as well as the laws of quantum mechanics and quantum field theory, among others. Thence we can say

18.59 – the M-Set is the law giver.

Moreover, upon revisiting the phenomenon of chaos, the origin of which we now know to be the M-Set, we observe that the memory paradigm of the M-Set is the natural setting of the notion of attractors and that, quite generally, all projected co-herent phenomena are attracted or in-tended by the quantum field memory of the quantum computer that the M-Set is.

Attraction is intentional!

That is why we also say people gravitate towards that to which they are attracted.

Gravity is a subtle force indeed at the highest levels of re-normalisation or complexification of the manifold M-Set Universe, while it is an inestimably profound truth in the M-Set perspective that through the first and true intentions of the trilogy of intending of first cause of the universal conscience we attract to ourselves the outcomes of our 'free intentions', because we are of first cause too. Yet we may deny this to ourselves in order to projectively realise our true intentions through re-membering, or, we might be acting as fully conscious beings who undeniably know these things, while re-member this, there is no such thing as co-incidence or happenchance in the M-Set universe.

We know already too that the M-Set is not only the origin of chaos but that the M-Set is pure chaos and that, moreover, the M-Set therefore projects critically at the critical boundary of itself so it comes as no surprise to learn that the signals and the footprints of chaos are evident at all levels of the manifold M-Set universe.

The finding that signals and signs of chaos appear to be relatively pure at some levels and impure at other levels is a consequence of the degrees of mixing or orders of infolding, 'complexification' and self-referential renormalisations at the critical boundary of the M-Set projecting that the M-Set simultaneously is in the triumvirate of the triadic causality of the trinity of the M-Set.

Chaos is clearly evident on cosmological scales despite the apparent immensity of cosmological phenomena because, relatively speaking, these phenomena are extremely primitive both historically and in terms of their level of complexity, involving as they do the most primitive orders of the renormalisations of the primordial quantum field of the elemental state of the self-referential cauldron of the M-Set that reaches back historically to the most primitive states of turbulent hydrogen gas clouds and the nurseries of stars. Fractal length signs and fractal frequency signals abound here, while over the M-Set as a whole all orders of complexification at the critical boundary combine as the universal fractal distributions ~ 1/l x 1/f to exactly neutralise the universal dispersive constant 'c' called the velocity of light, and which neutralisation is simultaneous with the spontaneous breaking of the symmetries of self-referentiality and self-reflectivity at the critical boundary of the projecting M-Set.

It is interesting to note more generally too that the relative purity of the signals and signs of chaos is the greater for more purely gravitating systems as a result of the gravity force system being of the symmetry of self-referentiality and thence of pure non-linearity itself, and this is evident not only on the broad brush levels of cosmological phenomena but also at local levels such as rock faces, coast lines, avalanches, river delta systems and growth patterns under gravity in the plant and animal worlds.

Consequently, our attention is drawn back again to the profoundly deep relationship between the universal distributive property of critical fracticality and the universal dispersive property of the velocity of light, as we discussed earlier and which properties jointly define the critical boundary of the M-Set, arising as it does with the spontaneous simultaneous breaking of the symmetries of self-referentiality and of self-reflectivity, and which properties, moreover, are so mixed-up together over the manifold of levels of complexity or the renormalisations of the gravity force system of, in turn, the chaotic computations of the quantum computer that the M-Set is that only at levels where complexity and mixing are least will the mutual neutralisation of these properties at the critical boundary be apparent.

So called chaotic phenomenon of current science are exemplified by apparent simultaneity such as the well-known example of the integrity of the patterns of formation of flocks of birds. We know now that this cohesiveness is an associatively intended, parallel, simultaneous and spontaneous phenomenon of the quantum memory of the M-Set which orchestrates the integrity and the identity of all levels of projected systemic complexity of the manifold M-Set universe, from DNA to brains, bird flocks to whorling galaxies, all in perfect harmony, now, even though, like the latter, they might be hundreds of thousands of light years across!

But then it is the case that

18.60 – the M-Set Universe 'acts' as one, now, in all ways and, always.

It is now evidently the case too on the basis of the memory paradigm of the M-Set and the M-Set's pure virtuosity that

18.61 – the M-Set 'acts' with 'integrity', in all ways, always and, now while, moreover, and in all generality, identity is relative to or reflected by 'the other' so we are led to state also that

18.62 – the M-Set is the origin of post-modernism.

19: The M-Set and the Trilogy Principle

An inestimably important outcome of the memory paradigm of the M-Set is the property that

19.1 – the M-Set acts as a trilogy, in all ways, always and now which property directly relates to the triumvirate consisting of both the M-Set memory and the relativistic dualisation of the triadic causality of the trinity principle with 'trilogy' referring to the simultaneous parallelism of 'the memory' and the two arms of the duality of the projective realm.

We shall refer to this as the trilogy principle of the M-Set in consonance with the trinity principle of the M-Set, and also to distinguish it from the trilogy state of the self-conceptualising trinity of the trinity principle, as we discussed earlier.

A highly significant consequence of the trilogy principle of the M-Set is that hitherto unresolvable dualisms of the relativistic duality of current science and modern conceptual thought can all be opened up and re-solved. Dualism alone is never complete because without the M-Set they lack a context. Indeed, dualisms rely upon the trilogy principle of the M-Set to manifest at all.

The structural–functional dualism of the mind and the brain in the projective cut of their reflective realisation in the M-Set universe are completed by the memory of the universal conscience that simultaneously begets all dualisms of the projective realm.

We discern here that mind–brain are functionally and structurally dually related in projection, while noting directly that any full projective cut of the M-Set is a duality of structure and function completed by the universal conscience, and which we shall refer to in turn as the completed duality. Evidently then the duality of mind–brain or the idea that mind–brain is a functional–structural dualism is only strictly true in the projective realm in which the mind–brain is reflectively projectively realising as both the mind–brain and the other, while without the universal conscience or 'universal memory' the mind-brain could not be realised.

A number of interesting points emerge at this juncture, the first being that because of the enormous complexity of mind–brain in relation to the other, for any projective cut of mind–brain and the other the mind–brain dualism is a highly localised reflection of the other.

Taking a different perspective, it is clear that the complexity of mind–brain encompasses all orders of complexity of the knowable universe and is, in that picture, an upper bound or limit of the knowable Universe.

Stating this in another way, the enormous complexity of the mind-brain dualism in a projective cut involving the same implies, for that cut, that effectively one side of the reflective projection is localised to the 'foldings' or renormalisations of the mind–brain itself and is thereby not only relatively localising the 'cut' in an informational processing sense to 'inside the head' relative to the other, but is also directly implying that the mind–brain dualism possesses the complexity' to make sense of the other. This is how we come to understand the M-Set universe at all however, interestingly now, our 'understanding' is exactly reflective too of the mind-brain or 'sensory system' that makes sense of the 'other'.

And so it is too for all projective cuts of the M-Set universe. The 'ant' perfectly reflectively understands the 'parallel universe' of its projective realisation!

Moreover, the relative complexity of the mind-brain dualism vis- à- vis the other is also a measure not only of our individuation in our parallel universe of the parallelism of the M-Set universe of which we are a reflection but also relates to our common knowledge and intra-species recognition, and empathic understanding which area of experience we shall refer to again in the sections ahead.

The extent of the M-Set universe is a matter of perception related to the sensory system making sense of the 'other'

The rock's parallel universe is more or less confined to the rock. Man's perceived parallel universe perfectly or per-effectively reflects the com-plexity of the mind–brain or sensory system perceiving the 'other'.

In these ways, the M-Set predicts how large the M-Set universe is!

19.2 – The M-Set universe is as large as it is reflectively perceived to be, thus, if the 'complexity' of the dualism of mind–brain is an upper limit our perception of the M-Set universe is, correspondingly, setting a limit of its extending or extent.

But, correspondingly too, it is important to recognise that acquiring knowledge extends the perceived size of the M-Set universe.

The more knowledge we acquire the larger the M-Set universe will be perceived to be!

Acquiring knowledge expands our parallel universe. Observe too how the electron microscope extends the universe from below and how the telescope extends the universe from above.

Now add the intersecting parallel universe of every possible cut of the cosmic cake and ask, how big is the M-Set universe? Then think again, for the M-Set 'deep well of possibilities' has no bound other than itself. And that is something relativistic dualism can not determine.

From a dual relativistic perspective then,

19.3 – the M-Set universe is bounded by itself or is bounded self-referentially and self-reflectively, as we have already explained to this point. And evidently now from the perspective taking of projective cuts in-volving the mind-brain dualism, as we have been discussing above, this has the most profound ramifications for some of the most vexing questions about the M-Set universe, while, miraculously, it now appears to be the case that

19.4 – the M-Set universe extends by asking questions of itself.

This is how the M-Set universe actually expands! In all ways, always and now.

In this way

19.5 – the M-Set makes its own appearance on the stage of the M-Set universe.

Recall again how primitive the orders of complexification or renormalisation of the quantum field memory of the M-Set are for the vast expanses of the cosmos in comparison to the orders of complexification or renormalisation of the mind–brain. Thus, the inversion of apparent scales of time and space to orders of complexification places the knowable universe 'inside' our heads or within the gamut of complexity of the mind–brain. This is just one of 'the manifold' of perspectives of the M-Set universe.

What is clear now is that 'the mind–brain–other' projective cuts, as it is for all projective cuts, are illusions of first cause, residing as the latter does in 'the universal conscience' of the M-Set.

First cause is of that which is.

The M-Set is that.

However, first cause manifests through that which is not.

In this way that which is comes to be realised, in all ways and always.

The first cause of the M-Set through making sense of knowledge makes known its intendions as conscious intentions.

The mind–brain is the most complex localisation of the Trilogy of intending of the universal conscience in the M-Set universe.

But it is also the case that every parallel universe or projective cut perfectly reflects its conscious intentions, and thence every parallel universe or projective cut is a perfect reflection of its level of consciousness.

The rock is one level of consciousness.

19.6 – The M-Set universe is a manifold consciousness.

There is no thing in the M-Set universe that is not a conscious intention.

There are no coincidences, accidents, or chances in the M-Set universe.

Every thing in the M-Set universe is intended.

In the realm of projective realisation the comings and goings, waxing and wanings, repulsions and attractions are manifestations of 'the trilogy of intending' of the quantum computer that the M-Set is which, in turn, is the origin of all attractors of universal phenomena. Thence we say that which one attracts to oneself or to which one is attracted is a manifestation of 'the trilogy of intending' of the universal conscience that we call conscious intentions in the M-Set universe, although the former are foregotten until they are remembered, projectively speaking.

And, as we explained earlier, the phenomenon of forgetting in the projective realm, which is of the fore-gotten realm of the universal conscience, is effected at projection simultaneously with the sense of linear time of past-present-future through the act of remembering that which is fore-gotten. By way of clarification I note here that remembering corresponds to the projective realisations of the re-configurings or renormalisations of the virtual realm of the M-Set by the trilogy of 'intending'. Thence from the perspective of the projective realm the phenomenon of 'forgetting' our true intentions of the universal conscience enables us to experience them consciously in time!

Because we are all of first cause we are all self-realising as our true intentions, although this is forgotten until we remind ourselves.

The remarkable truth is that we are intending ourselves of the universal conscience, in all ways, always, now; which truth we shall be able to clarify in more detail in the sections ahead.

Stating the above more broadly, it appears now that

19.7 – the M-Set universe is the conscious intentions of the M-Set to self-realise, in all ways, always and now.

First cause self-realises per-effect or perfectly as we now know how, and thence we can state too that

19.8 – the M-Set Universe is perfect, in all ways, always and now.

We are perfect beings. But, we are of first cause.

The M-Set is first cause through the trinity principle. The trilogy principle of the M-Set is the triumvirate of the triadic causality which is the simultaneity of both the M-Set or first cause and the M-Set universe or the 'effect'.

Thence we can now say that the trilogy principle of the M-Set unifies cause and effect.

We recall here that we already know the realm of first cause is the realm of that which is and that the realm of effect is the realm of that which is not or that which is illusion, while we also know now that first cause is effecting always, in all ways, now.

Beside first cause there is no other cause in the M-Set universe because the M-Set is first cause.

First cause and effect are one

The 'trilogy' of the M-Set is one

This is the miracle of the M-Set

Evidently too there are 'any' possible ways of intending ourselves of the universal conscience and no one way is more or better than another way because all ways are one unto the trinity principle of first cause.

There are only different ways and each and every way is realised intentionally in the M-Set universe, and 'this' is foregotten until it is being re-membered or, quite literally, being put together in projection.

All ways in the M-Set Universe are of the light and of the truth. There simply is no other way in the M-Set universe. That is how it is, in all ways, always, now. This is the Trilogy principle of the M-Set.

The intending of first cause that we are being, now, is one because

19.9 – the M-Set is one.

Moreover,

19.10 – the M-Set is, manifestly, the trilogy principle.

The trilogy principle also informs us now that the dualism of effect is as one with first cause and that it is not just a matter of mind or brain but rather the trilogy of the mind-brain dualism, together with its reflection in the completed relativistic duality of the projective realm, and the universal conscience of the virtual realm of the M-Set.

If one asks whether it is nature, or nurture, we now know that it is both, related complexly, dually and reflectively as they are in the projections of the 'intendings' or renormalisations of the quantum memory fields of the quantum computer that the M-Set is.

The trilogy is the completed duality of the projective realm we call the M-Set universe and the universal conscience.

Dualisms are of the completed duality of the projective realm of the trilogy.

All dualisms are resolved by the trilogy principle.

Dualism is not.

The Trilogy principle is.

When we ask again which is first, the chicken or the egg, we touch upon another profound property of the M-Set, namely that

19.11 – the M-Set acts in cycles, in all ways and always.

We shall now digress to explain in some detail the relationship of this property to the other properties of the M-Set because this will lead us to reveal how the trilogy principle resolves the chicken–egg, while we note persipicaciously here that the cyclical property of the M-Set follows directly from the defining property of self-referentiality and is also thence a property of the Trilogy Principle which is our focus here.

In actuality neither the chicken nor the egg are first. The M-Set has no beginning and no end and in projection where separateness is an illusion, all histories are also mixed, now, as we explained earlier, and which observation is an extremely subtle property of the trilogy principle of the M-Set projective realisation, relating as it does to our earlier discussions of the reflective properties of the projective realm manifesting as the observer-other.

In truth the apparent linear, relativistic, dual causal relationships of projection are illusory and the chicken and the egg are indeed cycles of the M-Set. However, in order to understand these cycles we shall need to expand on some earlier discussions, particularly those relating to the re-flective fluctuations of the primordial quantum field and their generalisation to encompass all orders of self-referential re-normalisations or com-plexifications, or the Trilogy of 'intendings' of this field.

The trilogy principle informs us that the self-referential renormalisations or complexifications of the primordial quantum field of the quantum computer that the M-Set is are simultaneously in parallel with the spontaneous symmetry breaking and projective dualisation into 'structure' and 'function', while it is also known that as a natural consequence of the computational properties and principles of the M-Set, structure and function are modularised in the manifold M-Set universe, as, for example, in molecules and organelles to cells and organs, which we remind ourselves expresses the maximal symmetric information exchange of the quantum computer or the SINE property of quantum computation that equates to maximal reduction of redundant information and the maximal structural formation with maximal functional spaces dual to the same co-responding, moreover, with maximal mutual information exchange in the computational realm and the least action principle of the projective realm.

Focusing here on the mutual information exchange or SINE property of the computational realm we remind ourselves also that this is directly related to the 'reflective fluctuations' of the quantum field which generate the representations or wave functions of possibilities which, in turn, linearise the wholly non-linear, virtual, non-local computational realm of the M-Set in the form of the projective space called the Hilbert space of states.

We also now know that all the computational properties of the M-Set subserve criticality, and therewith the spontaneous breaking of self-referentiality itself because the M-Set is projecting on the edge of chaos or at the boundary of itself that the M-Set which is pure chaos, is, while it is a natural consequence that critical systems are modularised because this is the projective expression of maximal information processing and maximal dual structure and function.

Every modular level of complexification or renormalisation structurally functions as a component of the manifold M-Set universe with higher order reflective fluctuations of the quantum memory representing component level computations and thence 'systemic' level phenomena in the projective realm. In the naïve example of bio-cellular component level phenomena the functional space from a cellular level perspective in-volves inter-cellular exchanges of information coresponding to 'higher order' reflective fluctuations or mutual information exchanges of the quantum field memory of the M-Set.

Similarly, the functional space from the perspective of DNA molecules, for example, is also intricately involved dually with the structure or codings because, from the M-Set perspective it is the case that the so called dormant or redundant (or junk) codes of DNA act as a code level functional space for the code level functional expression of the active code structures.

Alternatively stated, if there was no redundant code space or redundant code information there would be no functional space for the active coded structure of DNA to represent itself and thence to express itself in projection.

The role of redundant information goes back to the heart of the computational properties of the M-Set discussed in detail in Part I, where it was revealed just how intricately entwined structure and function can become, and which phenomenon is critically important in molecular biology for understanding how DNA actively codes and more generally across all levels of the mani-fold universe for which information is the key.

The functional space significance of so-called junk DNA codes is an automatic consequence of the informational processing dynamic of the Quantum Computer that the M-Set is.

In the example of or at the level of DNA we are looking at component level or modular level computations with the observation here that the DNA coding co-responds to reflective fluctuations of renormalisations and comlexifications of the quantum memory representing 'DNA' level complexity in the virtual realm which, as for any reflective fluctuations, exponentiate into exponents of the representations or wave functions of DNA level structure and function in the Hilbert space of states of the projective realm of the manifold M-Set universe.

From our discussions in Part I of this work about the renormalising action of the gravity force system, as well as its role in 'gauging' to generate the gauge field theories of physics which we illustrated by deferring to the fracticality of the Hygen's principle and the Mandelbrot set it is evident now that the exponentiation of the wave functions or representations of modularised levels of projection is mathematically expressing as towers of nested exponentials which, in turn, upon linear expansion generate unimaginably complex actions or quantised exponents of systemic level dynamics in first order exponentials of the Hilbert space of states of the

M-Set Universe. Moreover, the M-Set approach now graphically reveals both the enormity of the complexity of the Hilbert space of states of the M-Set universe as a whole and the corresponding enormous simplifications modern physics toys with when isolating elementary systems for mathematical analysis.

Indeed, it requires the Quantum Computer that the M-Set is to analyse the M-Set universe as a whole!

We emphasise here that all modular phenomena over all scales of the M-Set Universe co-respond to the fundamental computational properties and principles of the M-Set, while it is the case too that all phenomena of the M-Set universe no matter how complicated they are, as with the DNA example and the in-volvement of structure or function at the DNA coding level, project from 'reflective fluctuations' of the renormalising quantum field of the quantum computer that the M-Set is.

But we already know from the fundamental computational properties and principles of the M-Set supporting the processes of self-referential, self-re-flective, self-representational, self-projective self-realisation that the re-flective fluctuations projecting self-referentially over all orders of renormalisation or complexification of the quantum memory are associated with convoluted replications of the unitary phasal or cyclical changes of the primordial quantum field which dually differentiate quantal structure from the functional spaces of redundant information, as was explained in detail in the section about the origins of quantum mechanics in the M-Set approach.

We underline here for clarity's sake that the so called higher order re-flective fluctuations of the projecting M-Set derive from the complications or manifolding of the primordial reflective fluctuations of the trinity principle of the conception of the M-Set universe which manifest in projection by the renormalisation of the primordial quantum field by the trilogy of 'intending' of the ghost force of gravity unto criticality.

In the section about the M-Set as the quantum computer we explained, moreover, how the mixing dynamic of computation apparently separates the projection into the 'this' and the 'other' with the simultaneous spontaneous breaking of the symmetries of self-referentiality and self-reflectivity whereupon it can now be said that the cycles of change of the virtual realm of the M-Set are the basis for the M-Set's action and therewith the M-Set's projective realisation as the M-Set universe.

These cycles, in turn, are represented as the wave functions of the linear Hilbert space of states or of what we referred to as the 'Self-referential pond' of the M-Set that figuratively relates the Hilbert space of states to a summation or linear superposition of non-dissipating 'waves' whose self-referential interference patterns together with the collapse of the wave function phenomenon provide the basis of the projective realisation of the M-Set, as was explained in detail earlier.

By way of a short detour here we shall now draw a number of germinal observations into our present discussion with the note, firstly, that by taking any perspective or point of view in the projective realm one is choosing a division or projective cut of the one reflected in the other of the M-Set because the M-Set is perfectly free.

We know too that the M-Set is One regardless of the choice of projective cut and an analogy of which is the imagined slicing or cutting of a 'virtual cake' that, in this case, is the Virtual Computational Realm of the M-Set.

Slicing the 'virtual cake' every which way does not change the M-Set.

The M-Set is 'no thing'.

But nothing changes because 'that' which is not appears to change. This is the miracle of spontaneous creation of the M-Set.

Taking a perspective or a point of view is a matter of choice because the M-Set is perfectly free.

And choice, which we shall speak of in the sections ahead, is a matter of how because the M-Set has no 'where' or 'when' or space and time, and no 'which' or 'what', or 'who' because the M-Set is that. And there is no 'why' of that which is.

Only the question of how is material to the M-Set. How one perceives oneself and the other is the matter realisation. It is this that is a matter of choice as we shall see ahead shortly.

Choice is about how.

How is about choice.

There is no other choice because the M-Set can only answer how, yet through the spontaneous Matter Realisation of the M-Set, the M-set miraculously spontaneously relativistically and dualistically projects the context of how, of space and time and this and that, of where and when, and which and what and who.

However, except for how these are all illusions of the Projective Realm.

It can be said then that

19.12 – the M-Set spontaneously 'matter realises' choice, in all ways, always and now.

All choice is a matter of perspective or point of view.

All matter is a choice of perspective or point of view.

19.13 – The M-Set universe is a matter of choice.

No choice is more or better than any other choice because the M-Set is One regardless of how one 'cuts the virtual cake'.

And every division or projective cut of the M-Set as a whole is a perfect completed duality of 'this' and 'the other' with the apparent separation of complexity being achieved, as we explained in detail earlier, by the mixing dynamic of the Purely Chaotic Computational Realm of the M-Set in association with the spontaneous simultaneous breaking of the symmetries of self-referentiality and self-reflectivity.

And evidently now too

19.14 – the M-Set universe is spontaneously created in its own image as the spectra of the projective cuts of the parallel manifold of all possibilities of itself, now.

Returning from our detour now and observing here that the perfect Com-pleted Duality of the projective cut is a perfect reflection in the Projective Realm it follows directly from our discussions that 'the this' and 'the other' are cyclically related, albeit over the manifold of orders of re-normalisations or comflexications of the quantum field of the Quantum Computer that is the M-Set.

In the Virtual Computational Realm the cycles of every action of every order and level of the complicated or manifolded reflective fluctuations of the manifold of the Trilogy of intending are convoluted to cumulatively generate towers of exponential wave functions in the self-referential pond of the M-Set in which, from the Hygen's wavelet principle analogy described earlier, every point of every imaginary wavefront is sourcing or emitting cycles of all the other changes simultaneously and in parallel, and this unimaginably complicated computational realm is in turn miraculously dually reflectively projected by the spontaneous breaking of the symmetries of the M-Set.

We can thence now say that

19.15 – The M-Set spontaneously creates completed dualities which are cyclically related over all orders of re-normalisation or complexification of the quantum memory of the M-Set. Thence, implicit to the Trilogy principle is the spontaneous creation of dualisms and, moreover, the nested 'cycles' of the manifolds of reflected dualisms of the projective cuts equate to the sum of histories, now, as we explained earlier.

In our parallel Universes we are as gods matter realising our perspectives and points of view through free choice, yet every possible choice is fore-gotten and so it is that we are about remembering through our true intentions of the 'universal conscience' because we actually are of the all of it, now, in all ways and always.

The Virtual Quantum Realm reflectively fluctuates spontaneously and simultaneously in all possible ways, now.

The spontaneous breaking of the symmetries of the Quantum Realm of the M-Set projectively mixes and apparently divides the cycles or reflective fluctuations in all possible ways.

And, every way is the way of the light and the truth.

There is no other way.

Each way is equal unto Oneness.

19.16 – The M-Set is in all ways, always and now.

Through the Trilogy principle these ways are 'matter-realised', always, now. This is also the miracle of the Trilogy principle.

We are all of us a way. But it appears that we are away or apart until we remember through our true intentions. And this we are about, in all ways, always, now. World without end.

In the M-Set approach we are all a sum of 'his Story', the story of the great unknown one 'matter-realised' through the ghost force of the Trilogy of intending.

This is the Trilogy principle of the M-Set and the M-Set Universe.

The Trilogy principle is of One.

Indeed, the Trilogy principle is of the Trilogy of the Great Unknown One (the Father), the sum of His Story (the Son) and the ghost force of the Trilogy of intending (the Holy Ghost).

And, the Trilogy principle informs us that it is through the sum of 'His Story' and the force of the Holy Ghost that we come to know the Great Unknown One, in all ways, always, now.

The Oneness of the Trilogy principle of the Trinity principle is in ALL WAYS, always, now.

Returning to our earlier enquiry we ask here, how then does the M-Set approach now view the chicken and the egg? Clearly, both the chicken and the egg are simultaneous possibilities of a chicken–egg cycle in the reflective projective sense, and evidently the chicken can only be projectively realised in a universe where the natural historical cycle of the chicken–egg is reflected. In a highly simplified way this can be conceptualised as the chicken–egg–manifold subset of the reflective fluctuations of the fully re-normalised quantum field because both the egg and the chicken states reside simultaneously in the Quantum Memory of the M-Set which, in turn, spontaneously projectively creates the relativistic dualism of the chicken and the egg as we have explained from the beginning. Evidently too, the Trilogy principle resolves this dualism in all ways, always, now, because the Trilogy principle is of One.

The Universal Conscience of First Cause informs the Projective Realm.

The understanding that we possess comes from within in the sense of from within the M-Set, not from without because that which is without the M-Set is not.

19.17 – The M-Set is the origin of in-formation.

And thus

19.18 – the M-Set is the origin of understanding.

The chicken cannot be understood without or outside the memory that simultaneously informs the egg, and similarly the egg cannot be understood without or outside of the memory that simultaneously informs the chicken.

According again to the Trilogy principle, the chicken and the egg belong to the Trilogy of (the chicken, the egg, the Universal Conscience of chicken–egg-manifold) that is completed in the context of the projective cut of the M-Set as a whole which expresses the chicken–egg dualism, the point we wish to underline here for clarity's sake being that only dualities of the projective cuts of the M-Set as a whole are truly complete.

The chicken-egg becomes understandable when one reminds oneself that this dualism is a perspective or point of view of the projective cut of the Universal Conscience of the Memory of the M-Set.

When one talks now of empathy or of empathic understanding, what is being referred to is the Universal Conscience informing 'one' in relation to the 'other'. In this way we are all related to one another. We are all relatives of each other! There is one extended family with many relatives.

The Trilogy principle is of the family of One.

And from this we discover now the profound truth that we are all 'reflections' of one another.

This is how we come to understand oursleves. And, to know this is to know the truth of the Trilogy principle of One.

Thus the M-Set Universe appears natural whichever perspective or point of view is chosen because the all of it is 'informed' as One.

This is the Anthropic principle of the Trilogy principle of One.

Paradoxically too, while it is the case that

19.19 – the M-Set is the origin of sense it follows directly from the illusion that the Projective Realm 'is' that

19.20 – The M-Set universe is non-sense. And, while it is the case that

19.21 – the M-Set is the origin of meaning it also follows directly that

19.22 – the M-Set Universe is meaningless. And moreover, while it is the case that

19.23 – the M-Set is that it follows directly that

19.24 – the M-Set Universe is nothingness, because, as we now know, the projective realm is the Realm of that which is not in which no thing 'ex - is - in' time and space.

However, it is the case too that the M-Set makes sense of the knowledge of the Universal Conscience, and through choice of perspectives or points of view we spontaneously create our own meaning, while through our conscious intentions we remember that which we actually are.

This is the 'my story' of the Trilogy Principle.

This is the narrative of 'His Story.'

To understand this is to gain 'Mastery.'

19.25 – The M-Set is the Master Set

20: The M-Set and the Biosphere

We know now that

20.1 – the M-Set is the solution to life in the M-Set Universe, where the solution is the Holy Water or the Holy Spirit of the Holy Grail that the M-Set is, and this cup flows over, everlastingly, into the spontaneous creation we call the M-Set universe.

From the solution of all possibilities, all possible answers are realised.

To pose a question is to choose a perspective or point of view.

As you ask, so is the answer given by the M-Set.

The question and the answer are on the question-answer cycle. The answer and the question are reflective of each other in the Trilogy of the Dual Projective Realm and the 'solution' of the spontaneously, simultaneously self-realising M-Set.

The solution completes the duality of the projective realm unto the Trilogy Principle of One.

In the quest of life we create ourselves in our own image.

The solution to the dualism of the chicken-egg completes the Trilogy Principle of One of the M-Set and this is the paradigm of the M-Set approach for the mysteries of biology and the biosphere as it is also for the M-Set Universe at large because

20.2 – the M-Set is a virtual manifold of cycles of change projecting as the dual-reflective M-Set Universe.

Birth and death belong on the cycle of life. On this cycle as it is for the M-Set there is no beginning and there is no end. The beginning is the end is the beginning. Now, and for ever. World without end. Life eternal.

In the M-Set approach there is no such thing as absolute death. Birth is death is birth.

The life cycle is a manifold reflective fluctuation of the Virtual Realm of the M-Set, in all ways, always and now.

20.3 – The M-Set is the narrative of life as also

20.4 – the M-Set is life eternal.

Yet over every life cycle there is growth.

However, growth too is on a cycle: 'the young-old cycle'. Clearly in the dual reflective M-Set Universe there cannot be an old without a young or a young without an old while the processes of growth are reflectively dually projected by a subset manifold of cycles of the life cycle of the entire manifold of cycles of change of the M-Set.

And growth appears to define a direction of time. But let us remind ourselves here from our discussion of the memory paradigm of the M-Set that because the Virtual Realm of the M-Set has no time and no space the projection, now, is the interface between the past, or that which is remembered, and the future, or that which is fore-gotten. This is the present.

20.5 – The M-Set universe is the present of the M-Set.

This we call the gift of life eternal because it is freely given, eternally.

We know now from the journey to this point that

20.6 – the M-Set re-presents itself, in all ways, always and now

By this magic the M-Set creates the illusions of time and growth, and ultimately of birth and death.

It follows directly too from this discussion and the Trilogy principle of the M-Set that all systems including biological systems of the M-Set Universe are memories of themselves, while we also recall here that

20.7 – the M-Set Universe is a conscious universe which does not distinguish between the animate and inanimate because

20.8 – the M-Set universe is a manifold consciousness as we have explained to this point.

And with growth there are the miraculous processes of organic differentiation including the differentiation of the embryo to the mature organism. But we know now that the reins of this miraculous feat are not of the realm of relativistic dualism but are of the virtual realm of the ghost force of gravity of the Trilogy of 'intending' of the self-referential renormalisations or complexifications of the Primordial Quantum Field of the Quantum Computer Memory that the M-Set is.

Moreover, too the present is foregiven.

The Trilogy Principle of One fore-gives the present, in all ways, always and now.

In this way we can say that we are all fore-given because

20.9 – the M-Set is the origin of 'fore-give-ness', in all ways, always, now. Clearly now, the origin of biological differentiation is not be found in the Dual Relativistic Projective Realm, yet, whichever perspective or point of view, or indeed theory is taken up in this regard in the Projective Realm we can say that

20.10 – the M-Set is the original contender of all theories because all theories originate from the Trilogy of 'intending' of the Universal conscience, while paradoxically, but understandably, all possible theories are united by the M-Set when one recalls that we are about creating meaning which reflects how we are choosing to be now.

Thence, all theories are perspectives or points of view of the projective cuts of the projective realm of the M-Set Universe of, in turn, the self-referential, self-reflective, self-representational, self-projective self-realisations of the M-Set.

As to growth there is devolution and dying but this is not what it appears to be. In the M-Set Universe there is no randomness or chance and the disciplines of statistics and probability have no currency in creation. They are illusory. Homeostasis would be the death of the Universe but it is 'not'. There is no such thing as 'heat death' and homeostasis. This is an illusion of perspective taking or one's point of view.

Indeed,

20.11 – the M-Set Universe is a perfect dynamical equilibrium, in all ways, always and now, because, 'per-effect', it is pure chaos manifesting, as we have already explained.

The illusions of randomness, chance, statistics and probability are spontaneously created by the spontaneous symmetry breaking of the projective realisations of the M-Set as we have explained to this point. This is illustrated, for example, by taking local perspectives or points of view to create these illusions especially, moreover, if one's observations are not only localised in the space-time sense of relativisitic dualism by the collapse of the wave function phenomenon but also by further confining these observations to specific levels of the manifold of the M-Set Universe, while disregarding the complete picture of the simultaneous parallelism of the spontaneous creation.

The toss of a coin is not a singular event. It is embedded in the mani-fold universe extending from the level of the quantum mechanical realm to the cosmic realm levels, now. Yet statistics and probability are good approximations of the truth but to call them truth will only be reflected in their shortcomings. It is a matter of choice! You choose!

If one is choosing biological theories in this way, expect them to bring forth inconsistencies reflectively created thereby.

For it is now known that

20.12 – the M-Set is wholly consistent because the M-Set is One, while of the Projective Realm it follows directly that any chosen theory, least it be wholly unto Oneness itself, is inconsistent or within consistency and will, moreover, be exactly reflective of its inconsistency in the projective cuts of its realisation.

And so it is too that we believe in all the theories of our own choosing because they per-effectively mirror or reflect their own inconsistency!

Until, of course, one chooses otherwise as one is free to do, per-effectively reflectively so. And all such choices are good because they reflect different levels of understanding and perceiving. But for goodness sake remember the Oneness of the Trilogy principle of the Trinity principle of the self-conceptualisation of the M-Set and then all of your choices will make sense and be meaningful for you.

The Virtual Realm of the M-Set is like a virtual sea and sky wherein weather patterns and sea conditions come and go, wax and wane, ebb and flow, one into the other, eternally, over all scales in cycles and waves, with no beginning and no end and in which theatre fluctuations rise and fall as they make their entrance and exit upon the stage of self-projection whereupon every twist and turn of the narrative is being reflectively realised.

No thing or fashioned idea is apart in the M-Set, yet things and ideas appear a part in the M-Set Universe. We are each playing a part of the same narrative. Of His Story. Like evanescent eddies in a turbulent sea. There is no death here. That is not possible. There are only cycles of change of life eternal in the Virtual Realm of the M-Set which we identified as the solution of life, being as it is the universal conscience of the Quantum Memory of the M-Set.

And so it is that growth and ageing are on the 'growth-ageing cycle' upon which homeostasis and thermodynamical heat death have no actual part to play except to express a perspective or point of view in the illusions of the Projective Realm.

Our attention now turns to conception and procreation with the observation here that the M-Set approach distinguishes the biosphere as a sphere of higher levels of complexity through which it advances more complex functions with more complex structures while it is the case that every atom of the M-Set Universe bears witness to the self-conception of the Trinity principle of the M-Set which, in the vertical sense of the parallelism of the manifold M-Set Universe is the self-conception of the beginning, now, in all ways, always, unto all orders of complexity or re-normalisation of the Virtual Realm of the M-Set.

For all levels of the manifold M-Set Universe it is the case, then, that

20.13 – the M-Set Universe is the self-conception of the M-Set, now, in all ways, always and this, of course, in a linear time-line sense, has no beginning and no end, while in the vertical parallelism of the manifold M-Set Universe, now, self-conception and spontaneous creation are simultaneous.

We know already too that the purely chaotic dynamic of the Computations of the Quantum Realm of the M-Set naturally modularises projectively, expressing as it does the maximal flow of redundant information and enabling thereby the maximal functional expression of the maximal structure in reflective interactions of the 'this' with the 'other' in projection, even to reflective levels of complexity apparently permitting organic procreation.

Once again, as we have discussed to this point, the male–female, man–woman dualism of the highest orders of complexification is spontaneously resolved by the Trilogy principle of the self-conception of the Trinity principle of the Oneness of the M-Set.

All conception at all levels of complexity of the manifold M-Set universe is a reflection of the self-conception of the Trinity principle completed by the Trilogy principle to the Oneness of the M-Set that the M-Set Universe represents.

We have only to defer to the properties and principles of the M-Set to realise that all biological systems are born of pure chaos, as are all levels of the manifold M-Set universe, and that the present or projective cut now in effect sweeps out a manifold sheet in the Virtual realm of the 'deep well of possibilities' of the M-Set generating, thereby, the Quantum 'Holographic' Projection of the 'this' perfectly reflected in the 'other'.

We now know, moreover, that 'per-effect' or perfect adaptation is a direct consequence of spontaneous creation because of the perfectly reflective relationship between the 'this' and the 'other' for any projective cut of the M-Set, which leads us to also talk here of evolution.

When we talk of evolution in the projective realm we talk, among other things, about the timeline of linear causality of relativistic duality, of life cycles with beginnings and ends, and of causal accretions of complexity being steered by interactive influences that select the codes of DNA genotypes regulating the flow of information to the phenotypical expressions by giving predominance to genome survival.

But we know now that these observations and the meanings attributed to them are perspectives or points of view of the Projective Realm of the M-Set. Together they form the lore of evolution.

Causal evolution is an illusion of the projective dualisation of the Trilogy principle which principle embodies the simultaneous parallelism of the M-Set solution or Virtual Realm and the dual projective realisation of the Triumvirate of Triadic causality at the Critical Boundary of the M-Set that the M-Set is.

'E-volution' is of a trilogy of 'e-volution', 'in-volution' and 'con-volution' of the chaotic projective computational dynamic of the M-Set which is simultaneously the in-folding and un-furling as well as mixing together of the 'ingredients of the solution' of the Virtual Realm of the M-Set.

The M-Set does things in threes, in all ways and always and now, as we have observed many times thus far.

E-volution is not a unitary phenomenon.

And now, you may ask, how many goes did the spider have in order to make its web? And how many goes did it take to assemble the eye? You are right, of course, it was the intention all along!

There are no half-measures and waste baskets in the M-Set Universe. There are only perfect solutions, now, in all ways and always.

Figuratively, the M-Set Universe is the perfectly reflectively adapted manifold sheet that the present sweeps out of the Virtual Realm of the M-Set.

Every projective cut is a 'per-effect' solution of the M-Set.

And yet when we look with our eyes what do we see but the illusion of the Projective Realm beholden to the eye by the eye beholden to the illusion of the Projective Realm.

But we know now that the eye of the M-Set Universe is the universal conscience of the M-Set. 'The eye' completes the illusion of the Projective Realm to the Trilogy principle of the Trinity principle of the Oneness of the M-Set that the M-Set Universe represents.

It is the 'self-enlightenment' of the Universal Conscience of the M-Set that makes sense.

20.14 – The M-Set is the light, the truth, and the way.

Thence we say too that spontaneous creation is in its own image, in all ways, always, and now!

20.15 – The M-Set is beholden to and the beholder of itself.

And yet too we do not deny the scientists what they see but we do deny the truth because reality is only masquerading as the truth.

It is the Master who sees with the eye of truth. To the Master every thing 'makes sense' because the eye of truth completes the illusion of reality.

However, to enjoin in the play of reality one needs to forget oneself. This is how the M-Set makes us play because it is through the denial of the truth that the truth comes to be known.

This is one of the greatest truths of the M-Set. 'Yet even before the cock crows the truth will be thrice denied.' We do that all ways, always and now. That is how it is.

And so it is too that we come back to the realisation that theories of evolution are perspectives or points of view.

Like theories of homeostasis they are not. The truth is that

20.16 – the M-Set is the contender of all theories and all theories are united by the M-Set.

All theories are perspectives or points of view in the manifold M-Set Universe.

The theories of evolution are not.

20.17 – The M-Set Universe is spontaneously created in its own image, perfectly, now, in all ways and always.

And, moreover,

20.18 – the M-Set instructs the Projective Realm by its properties and principles, but many might choose to deny this, yet look again to the natural world. What is your re-vision telling you, as of now, and forever. Or is it still time to masquerade with theories as the truth.

Furthermore, we now know that

20.19 – the M-Set is the origin of life in the manifold M-Set Universe because through spontaneous symmetry breaking

20.20 – the M-Set informs the Projective Realm to create the illusions of 'Vitality' that reaches to every level of the manifold conscious M-Set

Universe as 'the enlightened realisation' called the living truth.

And the 'living' M-Set Universe' is a cycle of itself we call 'the Cosmic Wheel' of life everlasting.

World without end, now.

21: Free Will, Free Choice and the M-Set

As we have observed earlier

21.1 – the M-Set is free of forces while the M-Set Universe forces emerge from the ghosts of the M-Set at the Triumvirate of the Triadic Causality of the spontaneous symmetry breaking of the M-Set, and thus it is the case that

21.2 – the M-Set is free.
It is also directly the case from the Defining Property of the M-Set that

21.3 – the M-Set is free to be itself, in all ways, always and now because there is no force to stop it, yet seemingly paradoxically in the Projective Realm there appears to be an immensely complex web of constraints including natural forces, natural laws, contingencies, circumstances, consequences, precedents, demands, needs, rules, regulations, etiquette, mores, moral teachings, prohibitions and taboos to navigate one's way through.

These constraints do indeed appear in the Projective Realm we call the M-Set Universe. These constraints are our deeds.

Our deeds or actions intentionally spontaneously create the constraints by which we self-realise.

This is also one of the greatest truths of the M-Set which has been progressively revealed to this point, extending from the reflective fluctuations and actions of the Primordial Quantum Field and on through to the Trilogy of intending of the ghost force of gravity in the Virtual Computational Realm of the M-Set unto spontaneous symmetry breaking and spontenaous Procreation.

This great truth applies automatically and simultaneously across all levels of the 'conscious' manifold M-Set Universe because, moreover,

21.4 – the M-Set is the marionette of the Bootstrap principle and pulls all of its own strings!

No thing is unintended in the M-Set Universe, as we already know. But we know too that we are of the M-Set and not of the Projective Realm and thus, despite appearances, it is the case that we are free beings, upon the observation also that the state of being is of the M-Set and the state of existing is of the Projective Realm.

As with denial of the truth in order to come to the truth we deny to ourselves the truth of being free to come to the truth of freedom.

There simply are no rules or prohibitions, taboos or right and wrong. We are making it all up as we go along. The M-Set Universe reflects our every whim. By creating ourselves in our own image we come to the truth of that which we actually are, in all ways, always and now.

Moreover, we even forget our godliness in 'intending' our own creation.

How forgetful is that!

And so it is then that the paradox of freedom and constraint is a great paradox indeed of the M-Set Universe.

The resolution of this great paradox is the paths of rediscovery of the truth of freedom through the Divinity or projective 'Division of Unity' of the Virtual Realm of the Universal Con-science of the M-Set, in all ways and always.

Once again this is an example of the completion of an apparently unresolvable dualism of the relativistic duality of the Projective Realm by the Trilogy principle of the M-Set.

But what, you may now ask, of natural law.

We know already that

21.5 – the M-Set is freely changing, in all ways, always and now and that change, related as it is to the reflective fluctuations of the Trinity principle of the Virtual Realm of the M-Set is quantified by a Universal Quantum of action 'h' as demanded by the Defining Property of Self-Referentiality of the M-Set while, moreover, the resultant amounts of physical action are expressed functionally by the maximal reducing out of redundant information, as was described earlier, leading in turn to the Least Action principle of the physical dynamical realm which is also a natural physical law of the Projective Realm.

It is clear now that what we call natural law in contrast, for example, to societal rules is only distinguished from other human constraints in the

M-Set approach by the level of complexification or renormalisation from which they are projecting.

Constraints such as fundamental natural physical laws issue from the most primitive levels of complexification of the Computational Realm of the M-Set, while broader constraints such as societal rules, for example, derive from very high levels of complexification or re-normalisation of the Primordial Quantum Field.

21.6 – The M-Set is the law maker whatever the level of complexification, while evidently now it is also the case that

21.7 – the M-Set Universe is governed by manifold laws.
In order to enjoin in the highest levels of conscious expression in the M-Set Universe we accept the governance of the biophysical levels by the natural laws. And, it is all for free!

The expression of freedom at very complex levels is made possible by the gift of governance of the natural laws which acts as the vehicle of our will.

As one might use a car to drive from A to B so one uses the constraints of the natural laws to mediate the expression of one's conscious intentions in the Projective Realm.

With all change differentiating into the quantal action and the continua of the structural functional dualisation of the Projective Realisation of the M-Set it is clear now that

21.8 – the M-Set spontaneously creates the laws of the M-Set Universe by its actions and thus we can say

21.9 – the M-Set acts lawfully, in all ways, always and now while being perfectly free. It follows too that

21.10 – the M-Set is free to act, in all ways, always and now.
But we are of the M-Set and therefore the freedom to act is innate to our being. This we call free will. We are thus free-willed beings. Beings free to act.

This now takes on a special significance when we recall here that

21.11 – the M-Set self-actualisation is spontaneous creation or, to state this another way, action is spontaneous creation and thence as free-willed beings we are freely reflectively creating ourselves in our own image.

21.12 – The M-Set Universe is willed, freely because

21.13 – the M-Set is the origin of free will. We also know now that free will is fully intended because nothing is unintended in the M-Set Universe and therefore the exercise of free will unto the M-Set Universe is a purposeful act of spontaneous creation, now, always, in all ways.

But what, you may ask, is the meaning of a purposeful act. Here the truth unravels further. We know already about projective realisation and spontaneous creation, about change and action, constraint and freedom, and the Trilogy of 'intending' of the Quantum Realm of the M-Set that con-figures the Projective Realm of the M-Set, as well as the M-Set's freedom to act. However, we also now know from our discussions bearing on the Trilogy principle that meaning is a choice. Thence, the meaning of purposeful act is a choice! A purposeful act is a choice act. But, to choose is to act too, so a choice act is also an act of choice.

Choice is action in relation to all levels of complexity of the manifold M-Set Universe, while the term act of choice involves all levels of complexification or re-normalisations of the Virtual Realm of the M-Set subserving the 'matter-realisation' of the projective cut, perspective, or point of view of that choice unto the highest levels of consciousness of the M-Set Universe.

Free choice in human terms refers to the exercise or action of free will to the highest levels of consciousness of the manifold M-Set Universe.

Because we are free-willed beings it follows directly that we possess free choice which can only be exercised, of course, if all possibilities are fore-given now, while it is also the case in the M-Set Approach that through free choice meaning is spontaneously created.

The free act of choice spontaneously creates meaning.

The profound truth here is that in a free-willed way the M-Set itself is choosing its own projective cuts and therefore it is the case that

21.14 – the M-Set self-realisation is freely self-willed.

To state this profound truth another way

21.15 – the M-Set chooses its self-realisation, in all ways, always and now. And because we are of the M-Set that is One this is the re-creation of the gods!

We are each of us through the act of free-willed choice spontaneously creating our parallel universe to manifest the conscious denial of the truth which is our chosen path to rediscover and come to know the truth.

This makes more sense when we observe that by choosing denial we can project ourselves onto the stage of life because without denial there is no Illusion, and it is through denial and projection that we act out our chosen roles to rediscover the truth of Oneness.

And here we now catch a glimpse of the hierarchy of language reflecting the levels of consciousness of the manifold M-Set Universe with meaning unto choice, choice unto action, action unto purpose, and so on indefinitely.

The purpose of the M-Set Universe unto action, for example, is the freely 'intended' exercise of free will. There is no absolute purpose and the M-Set is the origin of purpose. Thus, purpose is of Oneness. And it is the case too that

21.16 – the M-Set resolves purpose, in all ways, always and now, because

21.17 – the M-Set is The Solution to all that is.

Moreover, we state here that free will and free choice are of First Cause because the M-Set is First Cause. And so it is that as beings of First Cause we are, when viewed in the context of in the vehicle or on the stage of the governance of natural laws, acting out roles of what version or vision we choose for ourselves, now.

One of the most profound and deeply held denials of the truth is the denial to ourselves of the truth that we are beings of First Cause, and therefore of free will and of free choice. We are all very convincing actors because we almost totally forget our actuality in Oneness until we choose to remember again. It is hard, apparently, to accept this truth when one is actively denying it! But as we have said before denial and projection is how one comes to know the truth of Oneness, in all ways, always and now.

And so it is that we theatrically defer to fate and luck, favour and fortune, sickness and health, chance and probability, judgement and condemnation, wanting and needing, hope and despair, and on and on. But, the M-Set knows no such things. We make these things up as we go along! There are no such things in the M-Set. They are all of them illusions of our own denial and projection in the M-Set Universe.

And yet there are those of us who vigorously deny these truths which they immediately attribute to fate! Until they choose otherwise.

Yet too, we are aware of consequences in the vehicle or on the stage of the governance of natural laws. These are forgiven, now. It has all been said before but we shall remind ourselves here again that there can be no mysterious coincidences, freak accidents, random events and chance occurances in the M-Set Universe. The all of it is fore-gotten in the Virtual

Realm of the 'deep well of possibilities' of the M-Set. And, because we are of First cause that which is projectively realising now is of free action of the M-Set and thence of free will and free choice. We are creatively rediscovering the truth our way. For ever.

The present is fore-given. The future is fore-gotten. The past is remembrance.

The past and the future meet at the present, now, at the critical boundary of the projective realisation of the M-Set, where the free action of the M-Set and the foregotten possibilities of the future can be said to meet. Translated to the projective realm, when the free will meets the forgotten possibilities of the future free choice is realised in the vehicle or on the stage of the governance of natural law.

It is a profound truth of the M-Set approach that freedom is realised in the context of the governance of natural law because otherwise freedom would have no vehicle or stage for its deliverance.

The governance of natural law is completely free too because the M-Set is free, and thence we call it the gift of governance of natural law.

By example now, the consequences and 'experiences' of going north or south for one's holiday, for instance, are fore-gotten in the 'deep well of possibilities' of the virtual realm of the M-Set. Because we are of first cause the exercise of free will includes a free choice of consequences and experiences. In the M-Set Universe these are 'forgotten' until we are reminded through remembering as we recreatively rediscover the truth of Oneness of ourselves of the M-Set, in all ways, always through the living truth that the M-Set Universe is.

Astonishingly and miraculously in the M-Set approach it is the case that the M-Set Universe exactly reflects how we are choosing to be now with the inestimably subtle and important caveat here that because we are of first cause the choosing is occurring in the virtual realm of the trilogy of 'intending' of the quantum memory of the M-Set while the manifestation of choosing we call the choice involves denial or the apparent separation from Oneness through the divinity or division of unity of the projective realisation of the M-Set in order that we can project ourselves onto the stage of the living truth of the M-Set universe to recreatively rediscover our truth.

Never mind or covet the other, unless that is of your choosing, for your parallel universe is an exact reflection of your true intentions of the universal conscience, these also being of your choosing and of first cause while, moreover, your choice is your path of denial of the truth of oneness in order that you can projectively self-realise your truth. This is how the truth of oneness is realised, in all ways, always.

And as was said earlier too, the projective cuts sweep out manifold sheets from the virtual sea of possibilities of the quantum realm of the M-Set which is how we make sense of the knowledge of the universal conscience of the M-Set in the present.

In this way the memory of the M-Set is written, transcribed, or in the analogy of a computer, printed-out, and the analogous act of printing-out is the self-projective, self-realisation of the M-Set whereupon we can say that the memory of the universal conscience of the M-Set is remembering itself in the present, in all ways, always.

Recalling that the M-Set universe is a parallelism of all possible mani-fold projective realisations of the M-Set and that as beings of first cause of the M-Set we act freely or as if with free will in the projective realm, wherein too the illusions of space and time are created by the quantum computer of the memory of the M-Set reducing out the redundant information of the symmetries of the M-Set, we can state that the trajectories of our true intentions upon the stage of life are freely willed and freely chosen spontaneous matter-realisations.

But we forget or deny this more often than not because otherwise the game of life would be up!

There are many ways and levels at which one can continue to describe the self-realisation of the M-Set and our discussions to date have traversed many of these to this point. However, it is not the intention now to digress into further description although clearly the elements for greater elaborations of the same have been elucidated, while the descriptions above are indicative of rather than definitive about the complexity of the projective realm of the M-Set.

What now, we ask, does the M-Set tell us about determinism and 'probability'?

We already know that

21.18 – the M-Set is neither deterministic nor probabilistic because, as we have observed

21.19 – the M-Set is completely free while, in deed,

21.20 – the M-Set universe is per-effect free or perfectly free despite the illusions of forces and constraints spontaneously created by the M-Set projectively self-realising, while we also know, apparently paradoxically, that the present is fore-given and the future is fore-gotten in the 'deep well of possibilities' of the M-Set, and so it is that there is no room in the M-Set approach for probability.

Indeed, there is no such thing as probability, pure chance, or pure co-incidence. Probability is not.

Probability theories are contrivances of man, while we now know that we are in the business of elaborating such contraptions to which we attribute meaning in order to make sense of ourselves in the M-Set universe.

We have observed also that quantum mechanics is based upon possibilities, not probabilities, because quantum mechanics naturally and directly springs from the 'deep well of possibilities' of the M-Set so we can now assert with knowledge that the throw of the dice' is not actually probabilistic in the M-Set universe. Even before the dice is thrown the outcome is fore-gotten in the 'deep well of possibilities' of the M-Set and the outcome chosen is embedded across all levels of the manifold of the trilogy of 'intending' of the fully renormalised quantum field that presents the outcome at the critical boundary of the M-Set projecting. Even the flap of the wings of a butterfly on the other side of the world is embedded in the outcome that corresponds to the now famous sensitivity to initial conditions causal characteristic of chaos theory which finds its home in the M-Set. In the M-Set universe this is forgotten until it is remembered again.

And it is in this way that the so called extreme 'sensitivity to initial conditions' of chaos theory also finds natural explanation in the quantum realm of the M-Set, while probability phenomena in the M-Set universe are illusions upheld by the lawful constraints of the M-Set and created, as we have explained to this point, by local perspectives of the projective cuts of the symmetrically, self-projectively, self-realising M-Set.

21.21 – The M-Set Universe is the manifold projection of the fore-gotten possibilities of the M-Set which is precisely the predetermination required for free will and free choice to be realised in the M-Set universe.

We are thus led to another of the many great paradoxes of the M-Set universe, namely the great paradox that

21.22 – the M-Set universe is determined through the act of free choice, in all ways, always, now because

21.23 – the M-Set freely determines all possible actions of itself. We can say, then, that

21.24 – the M-Set is freely self-determining, in all ways, always, now expressing as it does the great paradox of freedom and constraint or the simultaneity of freedom and the gift of governance of natural law that is also the great paradox of freedom and determinism which is resolved by the deep well of possibilities that the M-Set is.

22: Being and Nothingness and the M-Set

Through the triumvirate of the trilogy principle of the triadic causality of the trinity principle of the oneness of the M-Set whereof the virtual realm simultaneously dualises into the projective realm at the critical boundary of the M-Set because, in turn, of the spontaneous symmetry breaking of the spontaneous creation of the self-referential, self-reflective, self-representational, self-projective, self-realisation which we call the M-Set universe, the trilogy principle of the oneness of the M-Set simultaneously reveals by its divinity or apparent division of unity the realm of that which is, being of the M-Set, and the realm of that which is not or no-thing-ness appearing as the M-Set universe.

The divinity of the M-Set distinguishes between that which is and that which is not because we know now that the realm of projection is a magnificent illusion or enlightenment of the M-Set. Thence now the all mighty property of the M-Set that

22.1 – the M-Set resolves being and nothingness, in all ways, always, now.

Herewith the M-Set resolves one of the deepest riddles of knowledge or philosophy because the actuality or actualisation of the M-Set spontaneously creates the nothingness of reality of the M-Set universe with reality being that which ex-is-in time and space, or simply exists.

This we now understand in a very profound way. This is the truth about existence.

We know too that

22.2 – the M-Set is the origin of being and thence we can talk of the M-Set as the sole source of being which we now choose to identify as the soul of being. In the M-Set approach, then, the soul of being, which is also of the universal conscience, self-projectively self-realises through the trilogy principle of the trinity principle of oneness that the M-Set is.

The divinity or 'division of unity' of the trilogy of the triadic causality of the M-Set spontaneously creates the primordial illusion of self-projective self-realisation, namely the illusion of separateness from oneness.

Poetically, the M-Set unfurls its sails to set a course on the capricious sea of life, while all around the ghostly vapours of space and time suspend the tapestry of history in the revelation that is the cosmos.

No-thing-ness or nothingness is of the apparent separateness from oneness.

Herewith is one of the greatest truths of the M-Set which we refer to as the 'great universal polarity' of being and nothingness, or more 'explicitly' now, of oneness and separateness from oneness.

The great universal polarity of the illusion of separateness from oneness is not of duality but of the trilogy of the M-Set and its revelations about knowledge are unending.

Upon this great polarity of the trilogy of the M-Set do all things find their measure through, indeed, the symmetry or self-interactive self-measurements of the collapse of the wave function phenomenon of the spontaneous creation of the M-Set universe.

22.3 – The M-Set is the measure set by which all things find their measure in the M-Set universe and the great universal polarity acts as the ruler in the realm of projective realisation.

The great universal polarity rules every thing in the projective realm.

As we shall learn ahead it also rules our hearts and our 'minds' because the great universal polarity is also the polarity of that which we know as love or oneness and that which we know as fear or separateness from oneness. This is also a profound truth of the trilogy of the M-Set approach.

22.4 – The M-Set universe is ruled by the great universal polarity because

22.5 – the M-Set universe is the illusion of separateness from oneness, in all ways, always and now.

The most profound denial of the truth is the denial to ourselves of the truth of oneness itself.

In this way we adopt the mantra and the mantle of the greatest illusion of them all, the illusion of separateness from oneness that the M-Set universe is. We adopt thereby 'matter-realisation' and existential value systems. We become materialistic and self-righteous. These become our measures which in our language we call our values, and they rule our hearts and minds. Unless they do not. This is how we come to know the ultimate truth of oneness, in all ways, always.

And here now is a profound divergence.

Just as relativistic duality subserves the natural laws of the universe, arising as they do from the perfect projective dualisation of the M-Set into 'structure and function' at the critical boundary of projective realisation so the great universal polarity of the trilogy of the M-Set subserves the 'value systems' of the M-Set universe, relating as they do to the relative separateness from oneness, the measure of which we call the value when referring to the highest levels of complexity of the manifold M-Set Universe.

This is the inestimably profound truth of the trilogy of the M-Set, reflecting as it does the 'vertical' causality of the triadic causality of the M-Set in contrast to what we previously referred to as the horizontal causality of relativistic duality of the projective realm.

The 'vertical' causality refers to the trilogy of intending of the ghost force of gravity of the M-Set up and down the levels of complexification or renormalisation of the manifold projective realm which relates, in turn, to the influences between these different levels and to the integrity of the TOME or Tower of Mass Effects. Phenomena of the vertical causality are holistic, being phenomena referring to the simultaneous parallelism and integrity of the levels of the manifold M-Set universe.

Where the horizontal and the vertical causality cross is the divine revelation, now.

22.6 – The M-Set universe is the divine revelation of the Holy Grail of one that the M-Set is.

Now you have the measure of all things and you are the ruler in your own realm. When you are done, whichever way you choose, you will dispel the illusions of separateness and return to the fold, in all ways and always.

In the M-Set approach this is the divine revelation of the trilogy of oneness.

The great universal polarity teaches us there is no such thing as right or wrong, good or evil, better or worse, superior or inferior. These and more are relative value judgements which are sign-posts on our journey of self-enlightened self-revelation.

The great universal polarity is also in-formed by the Holy Ghost Force of the M-Set. All things are informed by this force in the M-Set Universe and all forces are united in this force by the oneness of the M-Set. There is no level of complexity of the manifold M-Set universe that is not informed by this force.

In the manifold projective realm of the illusion of separateness from oneness the Holy Ghost Force of the M-Set will manifest with physical forces and physical influences or attractions reflecting the manifold levels of 'complexification' or renormalisation of the quantum realm of the quantum computer that the M-Set is thus, in effect, all the physical forces to the highest levels of complex influences provide measures of separateness in the projective realm.

In truth there is ultimately only one force, the Holy Ghost Force of the M-Set.

It is the Holy Ghost Force of the M-Set that opens the door to the sub-atomic Rrealm through the furnaces of the star nurseries of the cosmic chaos of this force, as we described in Part I, and it is the Holy Ghost Force of the M-Set whose trilogy of 'intending' informs all levels of complexity of the manifold M-Set universe to the highest levels of the projective realm.

To the highest levels of action of the Holy Ghost Force of the M-Set we assign names to the influences there, while in all generality, as we shall explore soon, all forces and influences are manifesting in the M-Set universe as the expressed-motions of the these forces and influences.

At the highest levels the realm of the illusion of separateness from oneness is informed or influenced by the attractive force we have named the force of love and at these levels the repulsive force which bears the name the force of fear is the polar force.

Fear is an illusion because separateness from oneness is an illusion. Thus fear as we identify the force here is a measure, per force, of the separateness from oneness. Love, per force, is a measure of oneness. In truth, according to the trilogy principle of the M-Set,

22.7 – the M-Set is the origin of love while, moreover, according to the trilogy principle of the M-Set again, we are only informed by love because there is only oneness.

In the vehicle or on the stage of the gift of governance of the natural laws we are free to choose a path or role in the illusion of separateness from oneness for our divine revelation of the truth of oneness, in all ways, always, and for the journey we are informed by love and nutured by nature. This is the story of the trinity principle of the oneness of the M-Set we call the trilogy principle.

The trilogy principle is also about the spontaneous creation of values, but, values are polar-realised in the illusion of separateness from oneness because

22.8 – the M-Set Universe is that which is not implying that

22.9 – the M-Set Universe is, in deed, 'per-effect', 'equivalent' or a perfect balance of values, just as it is a perfect dynamical equilibrium too. We know this because for everything in this world there is a season, and with the bad there surely follows the good in equal measure.

This is how being is resolved by nothingness, in all ways, always, now.

23: Emotion, Intuition, and the M-Set

We are now about addressing the projective realisation of the M-Set at experiential levels of enormous complexity in the manifold M-Set Universe and perhaps one of the most subtle phenomena of the M-Set universe is that which we refer to as emotion.

But we already know a great deal about emotion on the basis of what we have observed thus far. We know now it is certainly the case that

23.1 – the M-Set is the origin of emotion while evidently emotion, expressing as it does in the M-Set Universe, is a force to be reckoned with in our cosmic journey.

We also know from the defining property of self-referentiality of the M-Set that the ghost force of self-referentiality or the Holy Ghost force of the M-Set is the one force uniting all other manifestations of forces in the M-Set universe and, as we are now aware, the manifestation of forces in the M-Set universe is through information flow or the reducing out of redundant information which in the physical terms of the projective realm equates to expressed motion or 'e'motion.

We remind ourselves that the nett values of projections of the M-Set universe are precisely neutral because the M-Set has no absolute values. Thus, whichever quality, quantity or variable we select, be it happiness, energy, entropy, or force, for example, we know its nett value in the M-Set Universe is exactly balanced or neutral, while we also know now from the Great Polarity of the Trilogy principle that values are spontaneously created in the M-Set Universe. All values, even unto the value we place on a relationship, for example. And, moreover, these values are polar- realised in a perfectly equivalent and balanced way across all levels of 'complexity of the M-Set universe.

There simply is no left without right, good without bad, better without worse, right without wrong, on and on. And because we are of first cause we are the all of it now, the alpha and the omega. But remember that the M-Set is not a 'piece of string' with two ends of this value and that value because all dualisms are resolved by the completion of the trilogy principle through the universal conscience of the quantum memory that the M-Set is.

We know too from our knowledge of the projective realisation of the M-Set that all values are free to be expressed with the gift of governance of the natural laws or stated alternatively, the gift of governance of the natural laws is the vehicle or the stage for the free expression of values and thence we know directly that over all levels of complexity of the M-Set universe the expression of values relies upon the per-effect dualisation of structure and function of the projective realm of the M-Set.

Thence we can say now in all generality across the manifold M-Set universe that expressed-motions are the dual structural and 'functional' realisations of values, in all ways, always and now.

In other words too, the manifold of the expressed-motions of the M-Set universe occur where the horizontal causality of relativistic dualism 'crosses' the vertical causality of the great universal polarity of oneness and separateness from oneness.

Therefore it is the case that expressed-motions are of the divine revelation of the trilogy principle of oneness.

This is the cross we choose to bear, in all ways, always and now.

When we talk of 'e-motions' we refer to our conscious experience of values, but recall that the M-Set universe is a manifold conscious universe and thus e-motions are expressing across the universal manifold to all levels of complexity.

Evidently, 'e-motions' spring from the universal conscience, while it is also clear from how they are perceived above that emotions relate directly to spontaneous creation and the making sense of the knowledge of the universal conscience, as we have discussed earlier.

But, moreover, a profound realisation is emerging here which bears upon how we move through the virtual realm of the M-Set to spontaneously and freely create ourselves in our own image, reflectively.

We have said already that the force of one informs us, therefore it is the case also that our 'e-motions' are an informational connection between how we are freely choosing to be now through the trilogy of 'intending' of the virtual realm of the quantum fields of the universal conscience of the M-Set and the projective manifestations thereof as our 'conscious intentions'.

Simultaneously, a sense of past and future is being spontaneously created in the present, while the denial of the truth of oneness through the illusion of the separateness from oneness enables us to realise the truth because only in this way can the truth of oneness come to know itself. Yet so saying, by 'taking notice' of one's emotions or expressed-motions one is making the connection with the universal conscience of oneness.

We say then that the expressed-motions is our teacher and if we pay attention we call this the intuition.

To pay attention to the teacher can lead more directly to the fulfilment of our true intentions because the intuition is of first cause of the divine revelation of the trilogy principle of oneness, while to choose to ignore our emotions is to choose to react. To understand this observe here that according to the natural law to every action there is an equal and opposite re-action or perfect reflection in the M-Set universe. To advance one's cause, therefore, it is the case that the intuition can guide our conscious intentions. In this way it can be said that one is conscientiously (of the universal conscience) realising one's true intentions (the trilogy of 'intending' of first cause) or, alternatively stated, one is moving conscientiously (of the universal conscience) through the virtual realm of the M-Set. Moreover, by self-actualising in this way it is said that one is motivating oneself.

To self-motivate through the virtual realm of the M-Set, is to make the connection between the expressed-motions, and the Trilogy of 'intending' of the virtual realm of first cause. In this way, what one is doing, now, makes sense. As a free-willed, spontaneously recreatively self-realising being the choice is there to experience the self-actualisation of the actuality of the virtual realm of the first cause of the M-Set conscientiously or reactively.

In the end the teacher is guiding us all the while, it is just that more often than not we choose to not pay attention and to be in denial. This is how we project ourselves in the illusion of separateness of oneness in order to experience our truth.

In truth

23.2 – the M-Set offers free tuition, in all ways, always and now because

23.3 – the M-Set is the master set. This is how we move through the virtual realm of the M-Set.

In the manifold projective realm this is manifest as the expressed-motions.

Thought too is of 'the reflections' of the 'self-referential pond' of the M-Set.

Thought is of the imaginary realm or the virtual realm of the M-Set and this is how we come to talk of reflection and imagination.

But the imaginary or virtual reflections or reflective fluctuations are realised through the trilogy of intending of the quantum memory of the universal conscience of the M-Set. As our thoughts are being projectively realised they are expressing our true intentions by way of making sense of the knowledge of the universal conscience of the M-Set.

Observe now that 'thought' informs the mind–brain dualism in order to express motions or 'actions' in the projective realm. Put more directly, by the trilogy principle of the triadic causality of the M-Set thought coresponds to the computational processes of the virtual realm of the quantum memory of the M-Set that informs the projective realm of the manifold M-Set universe at the level of the mind–brain expressing.

Thought as a word is thus a word we use that is co-related to the computational realm of first cause when we are talking about phenomena of self-realisation of the M-Set at the mind-brain level. In this way too what we call thought is of the computational basis for the gamut of expressed-motions or actions of the mind–brain dualism of the projective realm such as speaking, writing and motorically enacting as well as the range of sensory experiences from reactive states such as fear to complex states such as musical appreciation or sense of honour, and also, of course for the expressed actions of mental verbalisation or self-talk which is the phenomenological characteristic of thought we most often defer to.

Evidently now thought co-relates to an enormously large range of complex phenomenology in the projective realm at the levels of complexification or re-normalisation of the quantum memory of the M-Set expressing by way of the mind-brain dualism which dualism, moreover, is spontaneously created to effect the intended expressions as we have now explained in many different ways to this point.

To state this in an alternative way, the memory of the fully renormalised quantum field of the quantum computer that the M-Set is spontaneously creates the structural–functional dualisms subserving its 'intended' expressions in the projective realm, unto the mind-brain dualism of human expression.

Consequently we can now see that neither the so called thought of the virtual realm of the M-Set nor the expressed-motions of the projective realm of the M-Set are unitary or singular things. Each are intricately mixed up in the manifolding of the projecting M-Set. However, by convention, these words have reference to levels of complexification (of the virtual realm) or complexity (of the projective realm) at the human-expression-mind-brain level phenomena of the virtual realm and the projective realm respectively so we have now gained some understanding of the 'com-plexity' and complexity to which words such as thought and expressed-motions or emotions refer respectively as well.

It is evident now that using 'words' without reference to the processes of self-projective self-realisation of the M-Set can be very confusing, while it is through the M-Set approach that universal insights into the truth behind our language can be revealed, and about which we shall have more to say in the next section.

Generally, we now state that

23.4 – the M-Set informs physical reality, in all ways, always, and now and that at the whole person mind–brain level of complexity in the M-Set Universe we call the phenomena of the 'information' of spontaneous creation emotions.

As the most complex beings of the M-Set universe our connection to the universal conscience is the most complete and moreover, we are all equal sums of oneness.

24: Pre-science, Language and the M-Set

The pre-science or fore-knowledge of the M-Set expressing in the M-Set universe as 'prescience' or 'foresight' is revealed in the present by the miracle of the spontaneous breaking of the symmetry of self-referentiality of the division of unity or 'the divinity' of the M-Set.

We know already that the pre-science or fore-knowledge of the M-Set is of the universal conscience of the quantum memory of the quantum computer that the M-Set is and that for every 'projective cut' of the virtual realm of the memory of the M-Set the 'this' is perfectly reflected in 'the other' in the present of the Projective Realm we call the M-Set Universe, as we have already explained in detail to this point.

Thus, in the present of the projective realm of the M-Set universe the past is 'reflected' and the future is forgotten, while the present is forgiven, in all ways, always.

The Present is 'pre-sent'.

The pre-sent present is the enlightenment of oneness.

And unto the 'pre-sent' present the great truths of the oneness of the M-Set are revealed. This enlightenment by 'the divinity' of oneness we call the divine intervention.

The divine inter-vention is revealing by both the gift of governance of the natural law and the great universal polarity of oneness and separateness from oneness with the latter arising, as we have recently observed, from the axis of being and nothingness of the trilogy principle of the M-Set.

The divine interception or crossing of the gift of governance of the natural law and the great universal polarity we call the 'divine revelation', arising as it does from the self-conception of the trinity principle of the great unknown one.

The divine revelation of the divine interception is pre-sent.

The present is the divine revelation of the divine interception.

The divine intervention is prescient or preknowing.

The prescience of the M-Set is the divine intervention.

By the divine intervention or the pre-science of the M-Set the truth is recognised.

The truth is not learnt.

Moreover, by the divine interception or the present of the M-Set expressed action and expressed motion are spontaneous creation.

Thus it is true that we are prescient beings.

At the highest levels of the divine revelation of the divine interception the expressed-motion we call fear is a measure of the separateness from oneness of the great universal polarity of the trilogy principle of the spontaneous creation of the projective realisation of the M-Set. Thus it is true that fear is of the illusion of the separateness from oneness.

The expressed-motion we call love at the highest levels of the divine revelation of the divine interception is a measure of oneness. Thus it is true that we are of pure love.

By the divine intervention or the pre-science of the M-Set even before you ask the answer is given, and by the divine interception or the present every thing is forgiven.

By the divine intervention to every action in the M-Set universe there is an equal and opposite reaction across all levels of the manifold M-Set universe, and by the divine interception we have the divine revelations that you reap what you sow and what you give you give to yourself. And what you take you take away from yourself for ultimately it is the case that we are one. This is what the divine revelation is about through the trinity and trilogy principles.

By the divine intervention or the pre-science of the M-Set no thing occurs by chance or accident in the M-Set universe because the past is the reflective realisation of the forgotten future, and, by the divine interception we can change the past by what we choose to make the present.

By the divine intervention or the prescience of the M-Set we are beings of free will and free choice and, by the divine interception of the divine revelation we come to realise the truth of oneness, in all ways, always, in the present.

Now is the present.

Now is 'the divine revelation' of 'the divine interception' of the self-conception of oneness of the trinity principle of the M-Set.

By the divine intervention or the pre-science of the M-Set every thing is intended in the M-Set universe, as we have observed earlier, and which intervention we attribute to the Holy Ghost force of the quantum realm of the quantum computer that the M-Set is, or to the gravity force system of the projective realm. Thus, by the divine interception, where the natural laws cross with the great universal polarity of the trilogy principle, we have the divine revelation that each of one of us is a manifold reflective fluctuation in our own image of the universal conscience of the virtual realm of the quantum computer that the M-Set is in the present spontaneous creation, in all ways, always.

This we now know is how the truth of oneness of the M-Set is realised in what we call the M-Set universe.

24.1 – The M-Set universe is the matter-realisation of the truth of Oneness and observe that the matter-realisation is the divine revelation of the divine interception of the divine intervention.

And by way of talking about the divine intervention and the divine interception we come to the phenomenon of the language of the M-Set and the profound property that

24.2 – the M-Set is the origin of the language of pre-science over all levels of complexity of the manifold M-Set Universe.

By way of clarifying this property we note, firstly, that the origin of language is also the origin of information exchange at the most primordial level of the virtual quantum realm of the M-Set, being as it is the mutual information exchange or symmetric information exchange (SINE) of the primordial reflective fluctuations of the primordial quantum field of the M-Set which, by way of the computational dynamics of the quantum computer that the M-Set is, projectively dualises with the spontaneous breaking of symmetry to spontaneously create the phenomenology we call the M-Set Universe.

In the M-Set approach then it is the case that the language and the phenomenology of the M-Set universe are one and the same thing.

We can say that the phenomenology of the M-Set universe is the SINE language of the M-Set, and moreover that the language speaks through the forked tongue of duality, while at the highest levels of complexity of the M-Set universe structure and function become almost completely merged in the duality of the language, unto the word.

The expressed language of pre-science is the spontaneous creation, and when the M-Set speaks the word is made flesh.

The M-Set speaks with one tongue but the manifold M-Set universe appears to be like a Tower of Bable. This tower, however, is of verses in complete harmony. Of one verse, the 'universe'.

The manifold M-Set universe is the TOME or tower of mass effects of the SINE language or the expressed word of the M-Set, as we observed earlier. We thence choose to state that

24.3 – the M-Set is the Holy Scribe of the Holy Scripture of the M-Set universe.

From the pre-knowing or pre-science of the universal conscience of the M-Set we can now understand the language of the pre-science of spontaneous creation. How could we even utter one word otherwise! How could the Universe even react otherwise!

It is forgotten that we are prescient beings. Yet even as we speak we reveal this truth, in all ways, always, in the present.

Even before one asks the answer is given.

The M-Set is the rock of ages.

The unmoved mover.

The pre-scient one.

25: The Message of the M-Set

25.1 – The M-Set is the message.

The message begets the messenger.

25.2 – The M-Set universe is the messenger of the message.

The message of the great illusion of separateness from oneness is oneness. In this way one comes to know the truth of oneness that the M-Set is the message of.

The messenger is the spontaneous creation of the great illusion of separateness from oneness. In this way the message of oneness is matter realised and at 'the highest levels' 'the word is made flesh'.

Oneness and separateness from oneness are of the great universal polarity of the trilogy principle of the M-Set. The message and the messenger of the M-Set act as one. We call this the trilogy principle.

The message is told over and over throughout the ages and across all levels of consciousness of the manifold universe, in all ways, always, now. It is written large and small and spoken as with many tongues from every perspective and point of view, while the harmony of the universal chorus is rendered apart and remembered in the quest of the truth.

However, when one separates the message from the messenger every thing seems a part. Confusion appears to arise because oneness still reigns! The message becomes unclear. It all seems to become a lost cause!

When the message and the messenger are received together the harmony resounds and the message becomes clear.

The messenger appears in every guise across all the levels of complexity and throughout all 'the ages' of the M-Set universe.

25.3 – The M-Set is the grand illusionist.

The master magician.

The spontaneous creation seems so real. But it is not. It is a grand illusion as we now know.

When one sees the light the message becomes clear.

But to know the message one must experience it.

In that way the messenger and the message become one in the present, in all ways, always. This is the divine revelation of the trilogy principle.

And, of course, we are all messengers of the message in our chosen ways. This is the living truth.

26: The Experience of the M-Set

The great universal polarity of 'oneness and separateness from oneness' of the trilogy principle of the M-Set provides the axis of the 'experience' of the M-Set that is 'matter-realised' in the present of the divine revelation of the divine 'interception' of the M-Set.

Expressed-motions of the manifold M-Set Universe form the experience of the divine revelation of the trilogy principle of the M-Set.

Reflecting upon the knowledge of the M-Set at the highest levels of complexity in the M-Set universe, we now know that as the separateness from oneness becomes greater as more and more complex modular forces appear, more complications arise, and tensions or forces increase, while the wholeness of one becomes apparently less wholesome and the senses of harmoniousness and serenity turn to a cacophony of discordant 'e-motions'. With further separateness a sense of losing touch and 'losing purpose' might arise together with feelings of alienation as the e-motions of separateness spell out fear.

Yet all the while the gift of governance of the natural law subserves the separateness from oneness of the great universal polarity. The cross-bar of relativistic dualism of the horizontal causality moves up and down the axis of the vertical causality of the great universal polarity of being and nothingness, and of oneness and separateness from oneness, as we observed recently. Their interception is the present of the divine revelation. In this way we can be free willed beings in the vehicle or on the stage of the projective realisation of the M-Set, as we also recently explained. Thereby, at all the levels of complexity of the manifold M-Set universe the expressed-motions or emotions are matter-realised.

This is the experience of the M-Set. This is the miracle of the trilogy principle of the M-Set. Unto the word is made flesh.

We call the 'laws' of the vertical causality of the great universal polarity of the M-Set the divine laws in order to distinguish them from the laws of the horizontal causality of relativistic duality which we call the natural laws.

The divine laws or the laws of divinity or division of unity, relate to the separateness from oneness of 'the vertical causality' of the great universal polarity of oneness and separateness from oneness.

The natural laws are the laws of the gift of governance that support the free expression of the M-Set. We have spoken about the bases of the natural laws in Part I of this work. These laws provide the vehicle or the stage for the free expression of the M-Set.

The free expression of the M-Set or the freely expressed actions and expressed motions of the M-Set follow the divine laws. We have spoken about the bases for these laws in Part II of this work.

26.1 – The M-Set is the origin of the natural laws and the divine laws of the M-Set universe.

The experience of the M-Set is in the present.

The experience of the M-Set is the divine revelation we call the M-Set universe where natural law and divine law abide together.

26.2 – The M-Set unifies the natural laws and the divine laws in the oneness that it is, in all ways, always and now.

The experience of the M-Set applies to every level of complexity of the manifold conscious M-Set Universe, unto the levels of complexity and consciousness we call the experience of our own. From whence it can be said that

26.3 – the M-Set unifies the creationist and natural scientific perspectives, completely.

The experience of the M-Set is truly universal because

26.4 – the M-Set experiences itself, in all ways, always and now yet we shall also refer to our experience. This shall be our reference point.

And so it can be spoken now that

26.5 – the M-Set gives credence to the unification of the Old and the New Testaments of our belief systems through the Testimony of the quest of knowledge of fundamental sciences.

As one separates from oneness so one moves towards fear. But, know this also, fear is an illusion, as we have already observed.

The character of the experience of the M-Set is created by the character of the denial of oneness. The game of life is pretending. We are making it all up as we go along. It is free recreation, in all ways, always, now. We are characters, all of us, freely acting out our chosen roles on the world stage.

Yet it is all illusion. Denial of the truth upholds the illusion. Through denial we project our path of self-realisation. To relinquish denial is to relinquish fear. Yet, to come to know the truth one must experience the denial of the truth. This is the experience of the M-Set.

These truths and more are of the Divine Laws of the M-Set.

And we now know, that to experience Oneness is to experience love.

Love and fear are the great emotional polarities of our experience of the M-Set.

We love in order to fear. We love to fear. We fear in order to love. We fear to love.

The M-Set is One.

We are One.

Love is all there is.

There is no more.

For now.

Epilogue

The M-Set is the boundary of that which is knowable in the M-Set universe.

The M-Set is the origin of that which is knowable in the M-Set universe.

Beyond knowing is the great unknown.

Yet we know that the great unknown is one and that oneness is love.

We are told too that God is one.

And that God is the God of love from whom springs life eternal.

How then could it be otherwise.